# Agribusiness
## Decisions and Dollars

### SECOND EDITION

# Agribusiness
## Decisions and Dollars

### SECOND EDITION

Jack Elliot

DELMAR
CENGAGE Learning™

Australia • Brazil • Japan • Korea • Mexico • Singapore • Spain • United Kingdom • United States

**Agribusiness: Decisions and Dollars, 2nd Edition**

Jack Elliot

Vice President, Career and Professional Editorial: Dave Garza

Director of Learning Solutions: Matthew Kane

Acquisitions Editor: David Rosenbaum

Managing Editor: Marah Bellegarde

Product Manager: Christina Gifford

Editorial Assistant: Scott Royael

Vice President, Career and Professional Marketing:
Jennifer McAvey

Marketing Director: Deborah Yarnell

Marketing Coordinator: Jonathan Sheehan

Production Director: Wendy Troeger

Production Manager: Mark Bernard

Content Project Manager: Kathryn B. Kucharek

Art Director: Dave Arsenault

Technology Project Manager:
Mary Colleen Liburdi

Production Technology Analyst: Tom Stover

For product information and technology assistance, contact us at
**Professional & Career Group Customer Support, 1-800-648-7450**

For permission to use material from this text or product,
submit all requests online at **cengage.com/permissions**.
Further permissions questions can be e-mailed to
**permissionrequest@cengage.com**.

Library of Congress Control Number: 2007940810

ISBN-13: 978-1-4283-1912-7

ISBN-10: 1-4283-1912-3

**Delmar**
5 Maxwell Drive
Clifton Park, NY 12065-2919
USA

Cengage Learning products are represented in Canada by Nelson Education, Ltd.

For your lifelong learning solutions, visit **delmar.cengage.com**

Visit our corporate website at **cengage.com**.

**Notice to the Reader**

Publisher does not warrant or guarantee any of the products described herein or perform any independent analysis in connection with any of the product information contained herein. Publisher does not assume, and expressly disclaims, any obligation to obtain and include information other than that provided to it by the manufacturer. The reader is expressly warned to consider and adopt all safety precautions that might be indicated by the activities described herein and to avoid all potential hazards. By following the instructions contained herein, the reader willingly assumes all risks in connection with such instructions. The publisher makes no representations or warranties of any kind, including but not limited to, the warranties of fitness for particular purpose or merchantability, nor are any such representations implied with respect to the material set forth herein, and the publisher takes no responsibility with respect to such material. The publisher shall not be liable for any special, consequential, or exemplary damages resulting, in whole or part, from the readers' use of, or reliance upon, this material.

Printed in Canada
1 2 3 4 5 6 7 12 11 10 09 08

# CONTENTS

**Preface** xv

**Acknowledgments** xix

**About the Author** xxi

## 1 Agribusiness Management and Marketing 1

OBJECTIVES 1
KEY TERMS 1
INTRODUCTION 2
MANAGEMENT 2
    Roles and Functions of Management 3
        Making Decisions 3
        Functions of Management 6
    Key Economic Principles of Free Enterprise 8
        Economic Opportunities of Agribusiness 10
        Government Policies and Regulations 11
        Human Resources 11
MARKETING 11
    Importance of Marketing 11
    Marketing Plans 12
    Competitive Environments and International Markets 14
    Types of Markets 14
    Managing Risk 15
SUMMARY 16
SUPERVISED EXPERIENCE 16
CAREER OVERVIEW 16
DISCUSSION QUESTIONS 17
SUGGESTED ACTIVITIES 17
GLOSSARY 17

## 2 Developing Personal Life Skills 19

OBJECTIVES 19
BUSINESS PROFILE 20
KEY TERMS 22
INTRODUCTION 22

PERSONAL FINANCIAL MANAGEMENT  23

    Reasons for Personal Financial Management  23

    Advantages and Disadvantages of Personal Financial Management  25

CHECKING ACCOUNTS  27

    Guidelines for Preparing and Completing a Check  29

    Reconciliation of Bank Statement  30

SAVINGS ACCOUNT  31

    Types  32

    Advantages and Disadvantages of Savings Accounts  33

ACCOUNT MANAGEMENT  34

SAVINGS AND INVESTMENTS  35

TIME VALUE OF MONEY  37

    Future Value: Compounding  37

    Present Value: Discounting  37

CREDIT  39

    Importance of Credit  40

    Sources of Credit  40

        Credit Cards  40

        Personal Loans  41

EMPLOYER EXPECTATIONS  42

    Appropriate Work Habits  42

    Personal and Occupational Safety Practices  43

SUMMARY  43

SUPERVISED EXPERIENCE  43

CAREER OVERVIEW  45

DISCUSSION QUESTIONS  45

SUGGESTED ACTIVITIES  45

GLOSSARY  46

**Inventory  49**

OBJECTIVES  49

KEY TERMS  49

BUSINESS PROFILE  50

INTRODUCTION  52

IMPORTANCE OF AN ACCURATE INVENTORY  52

    Statement of Owner Equity  53

    Income Statement  53

    Obtaining Credit  53

    Insurance  53

    Estate Planning  53

    Tax Management  54

WHEN TO INVENTORY  55

WHAT TO INVENTORY  56

Assets    56

Consumable Supplies    56

Capital Assets    57

INVENTORY VALUES    57

Determining the Inventory Value of Non-depreciable Assets    57

Determining the Inventory Values of Depreciable Assets    58

Methods of Depreciation    63

Straight-line Depreciation    63

Accelerated Cost Recovery System (ACRS)    63

Modified Accelerated Cost Recovery System (MACRS)    63

SUMMARY    63

SUPERVISED EXPERIENCE    63

CAREER OVERVIEW    64

DISCUSSION QUESTIONS    64

SUGGESTED ACTIVITIES    65

GLOSSARY    65

## Balance Sheet    67

OBJECTIVES    67

KEY TERMS    67

BUSINESS PROFILE    68

INTRODUCTION    69

PURPOSE OF A BALANCE SHEET    69

RELATIONSHIP WITH OTHER BASE FINANCIAL STATEMENTS    70

Feasibility    71

Financial Risk    71

Profitability    71

THE BALANCE SHEET ADDRESSES FIVE AREAS    72

Solvency    72

Liquidity    72

Risk-bearing Capacity    73

Collateral Identification    73

Business Trends    73

STRUCTURE AND COMPONENTS OF A BALANCE SHEET    74

Assets    74

Current Assets    74

Cash on Hand, Checking, and Savings    76

Cash Value for Bonds, Stocks, Life Insurance, and Other
Investment Vehicles    76

Notes and Accounts Receivable    76

Non-depreciable Inventory    77

Other    78

Non-current Assets    78

Other    78

Liabilities  78
    Current Liabilities  79
        Accounts and Notes Payable  79
        Current Portion of Non-current Debt  80
        Accrued Interest on Current Liabilities  80
        Accrued Interest on Non-current Liabilities  82
        Other  82
    Non-current Liabilities  82
        Notes and Chattel Mortgages  83
        Real Estate Mortgages/Contracts  84
        Other  84
        Deferred Tax Liabilities  84
Owner Equity  84
COMPLETED BALANCE SHEETS  84
    Liquidity Measures  87
        Working Capital  88
        Current Ratio  88
    Solvency Measures  89
        Debt-to-Asset Ratio  89
        Equity-to-Asset Ratio  89
        Debt-to-Equity Ratio  89
COST AND MARKET VALUATION OF ASSETS AND LIABILITIES  90
    Cost Valuation  91
    Market Valuation  91
SUMMARY  91
SUPERVISED EXPERIENCE  92
CAREER OVERVIEW  92
DISCUSSION QUESTIONS  93
SUGGESTED ACTIVITIES  94
GLOSSARY  94

**5**    **Income Statement  97**

OBJECTIVES  97
KEY TERMS  97
BUSINESS PROFILE  98
INTRODUCTION  100
IMPORTANCE OF AN INCOME STATEMENT  100
    Uses of an Income Statement  101
RELATIONSHIP WITH OTHER BASE FINANCIAL STATEMENTS  102
CASH AND ACCRUAL ACCOUNTING  103
REVENUE  106
    Income and Expense Summary Sheets  106
    Revenue Entries  109
    Accounts Receivable  110

EXPENSES 111

NET INCOME 114

    Net Income from Operations 114

    Net Business Income 115

SUMMARY 116

SUPERVISED EXPERIENCE 116

CAREER OVERVIEW 121

DISCUSSION QUESTIONS 121

SUGGESTED ACTIVITIES 122

GLOSSARY 122

## Statement of Cash Flows 125

OBJECTIVES 125

KEY TERMS 125

BUSINESS PROFILE 126

INTRODUCTION 128

STATEMENT OF CASH FLOWS VERSUS CASH FLOW STATEMENT 128

USES OF A STATEMENT OF CASH FLOWS 131

    Summary of Receipts and Disbursements 131

    Evaluating Past Financial Practices on a Historical Basis 131

    Projecting Cash Inflows and Outflows 132

    Measuring Liquidity 132

    Ability to Meet Financial Obligations 133

    Maintaining a Line of Credit 133

RELATIONSHIP WITH OTHER BASE FINANCIAL STATEMENTS 133

SOURCES OF CASH FLOW 134

    Operating Income and Expenses 134

    Investments 134

    Financial Activities 136

ADVANTAGES TO BUSINESS AND LENDERS 136

STATEMENT OF CASH FLOWS AND INCOME STATEMENT 138

    Statement of Cash Flows 139

    Income Statement 139

COMPLETING THE STATEMENT OF CASH FLOWS 139

    Code 1: Cash Withdrawals for Family Living 140

    Code 2: Cash Paid for Operating Activities 141

    Code 3: Cash Received from Operating Activities 142

    Code 4: Cash Paid for Investing Activities 142

    Code 5: Cash Received from Investing Activities 142

    Code 6: Distribution of Dividends, Capital, or Gifts 142

    Code 7: Cash Received from Equity Contributions 142

    Code 8: Principal Paid on Term Debt and Capital Leases 143

    Code 9: Operating Loans Received 144

    Code 10: Proceeds from Term Debt 146

    Code 11: Operating Debt Principal Payments 147

Other    148

SUMMARY    148

SUPERVISED EXPERIENCE    148

CAREER OVERVIEW    149

DISCUSSION QUESTIONS    149

SUGGESTED ACTIVITIES    150

GLOSSARY    150

## Statement of Owner Equity    151

OBJECTIVES    151

KEY TERMS    151

BUSINESS PROFILE    152

INTRODUCTION    154

PURPOSES OF A STATEMENT OF OWNER EQUITY    154

RELATIONSHIP WITH OTHER BASE FINANCIAL STATEMENTS    154

CHANGE IN TOTAL OWNER EQUITY    155

   Change in Retained Earnings    155

   Withdrawals    159

   Change in Contributed Capital    159

   Change in Asset Values    162

COMPARISON OF CALCULATED AND REPORTED OWNER EQUITY    165

SUMMARY    165

SUPERVISED EXPERIENCE    165

CAREER OVERVIEW    166

DISCUSSION QUESTIONS    166

SUGGESTED ACTIVITIES    167

GLOSSARY    167

## Analyzing Financial Performance    169

OBJECTIVES    169

KEY TERMS    169

BUSINESS PROFILE    170

INTRODUCTION    172

FIVE KEY AREAS OF FINANCIAL ANALYSIS    172

LIQUIDITY    173

   1.  Working Capital    173

   2.  Current Ratio    173

   3.  Cash Flow Coverage Ratio    176

SOLVENCY    177

   4.  Debt-to-Asset Ratio    177

   5.  Equity-to-Asset Ratio    177

   6.  Debt-to-Equity Ratio    177

PROFITABILITY   178

    7.  Return on Assets   178

    8.  Return on Equity   179

    9.  Operating Profit Margin Ratio   179

REPAYMENT CAPACITY   179

    10. Term Debt and Capital Lease Coverage Ratio   179

    11. Capital Replacement and Term Debt Repayment Margin   180

FINANCIAL EFFICIENCY   180

    12. Asset Turnover Ratio   181

    13. Operating Expenses Ratio   181

    14. Depreciation Expense Ratio   181

    15. Interest Expense Ratio   181

    16. Net Income from Operations Ratio   181

GUIDELINES FOR APPLYING THE USE OF FINANCIAL MEASURES   182

SUMMARY   182

SUPERVISED EXPERIENCE   182

CAREER OVERVIEW   184

DISCUSSION QUESTIONS   184

SUGGESTED ACTIVITIES   185

GLOSSARY   185

## 9   Planning and Decision Making   187

OBJECTIVES   187

KEY TERMS   187

BUSINESS PROFILE   188

INTRODUCTION   190

THE DECISION-MAKING PROCESS   190

PURPOSES OF A BUDGET   190

VARIABLE AND FIXED COSTS   190

BUDGETS   193

SYSTEMATIC PLANNING   196

CASH FLOW BUDGET   201

SUMMARY   202

SUPERVISED EXPERIENCE   202

CAREER OVERVIEW   203

DISCUSSION QUESTIONS   203

SUGGESTED ACTIVITIES   205

GLOSSARY   205

## 10   Business Borrowing and Investing   207

OBJECTIVES   207

KEY TERMS   207

BUSINESS PROFILE 208
INTRODUCTION 210
CAPITAL 210
BORROWING AND DECISION MAKING 212
TOOLS FOR EVALUATING CAPITAL NEEDS 213
    Measuring Leverage 215
    Evaluating Risk 215
    Maximizing Profits 215
    Satisfying Multiple Goals 217
    Capital Rationing 217
LOAN OPTIONS 218
    Loan Length 218
    Loan Use 218
    Loan Security 218
    Loan Repayment Schedule 219
SOURCES OF LOAN FUNDS 220
COST OF BORROWING 221
    Annual Percentage Rate 222
    Inflation 222
SUMMARY 222
SUPERVISED EXPERIENCE 224
CAREER OVERVIEW 225
DISCUSSION QUESTIONS 225
SUGGESTED ACTIVITIES 226
GLOSSARY 226

## Taxes 229

OBJECTIVES 229
KEY TERMS 229
BUSINESS PROFILE 230
INTRODUCTION 232
DOCUMENTS FILED WITH EMPLOYER 232
TAKE-HOME PAY 234
    Personal Records 236
FILING INCOME TAX RETURNS 236
BUSINESS TAXES 236
    Business Taxable Income 236
    Business Deductions 237
TAX MANAGEMENT 237
    Self-employment Tax 238
SUMMARY 238
SUPERVISED EXPERIENCE 238
CAREER OVERVIEW 239
DISCUSSION QUESTIONS 239

SUGGESTED ACTIVITIES  240
GLOSSARY  241

## Management Information System  243

OBJECTIVES  243
INTRODUCTION  244
INSTRUCTIONS FOR THE MANAGEMENT INFORMATION SYSTEM  244
    Record of Supervised Experience Program [Form 1]  244
    Exploratory Supervised Experiences [Forms 2–3]  244
        Planning Sheet [Form 2]  244
        Exploratory Supervised Experience Records [Form 3]  244
    Placement Experiences [Forms 4–6]  245
        Placement Experience Training Agreement [Form 4]  245
        Placement Experience Plan [Form 5]  245
        Placement Experience Evaluation Form [Form 6]  245
    School-based Agreement Forms [Forms 7–8]  245
    Improvement Activities [Form 9]  245
    Enterprise Budget [Form 10]  245
    Partial Budget [Form 11]  246
    Cash Flow Budget [Form 12]  246
    Inventory Forms [Forms 13–15]  246
        Non-depreciable Inventory [Form 13]  247
        Depreciable Inventories [Forms 14–15]  247
    Balance Sheet Statement [Forms 16–17]  247
    Record of Income and Expenses Code Sheet [Form 18]  247
    Record of Income and Expenses [Form 19]  248
    Income and Expense Summaries [Forms 20–22]  248
    Income Statements [Forms 23–24]  248
    Statement of Cash Flows Worksheet [Form 25]  248
    Statement of Cash Flows [Form 26]  248
    Cash Flow Statement [Form 27]  248
    Statement of Owner Equity [Form 28]  248
    Financial Worksheet [Form 29]  249
    Financial Ratios [Form 30]  249
    Trend Analyses [Form 31]  249
    Resume [Forms 32–34]  249
    The Diary [Form 35]  249
    Supervision of Student's Experience Program [Form 36]  249

## Forms

    Form 1:  Record of Supervised Experience Program  250
    Form 2:  Exploratory Supervised Experience Planning Sheet  252
    Form 3:  Exploratory Supervised Experience Records  253

Form 4:   Placement Experience Training Agreement   254

Form 5:   Placement Experience Plan   256

Form 6:   Placement Experience Evaluation Form   257

Form 7:   School-based Agreement Form A   258

Form 8:   School-based Agreement Form B   259

Form 9:   Improvement Activities   260

Form 10: Enterprise Budget   261

Form 11: Partial Budget   262

Form 12: Cash Flow Budget   263

Form 13: Non-depreciable Inventory   264

Form 14: Depreciable Inventory   265

Form 15: Depreciable Inventory   266

Form 16: Balance Sheet Statement   267

Form 17: Balance Sheet Statement   268

Form 18: Record of Income and Expenses Code Sheet   269

Form 19: Record of Income and Expenses   270

Form 20: Income and Expense Summary   276

Form 21: Income and Expense Summary   277

Form 22: Income and Expense Summary   278

Form 23: Income Statement: Cash   279

Form 24: Income Statement: Accrual   280

Form 25: Statement of Cash Flows Worksheet   281

Form 26: Statement of Cash Flows   282

Form 27: Cash Flow Statement   283

Form 28: Statement of Owner Equity   284

Form 29: Financial Worksheet   285

Form 30: Financial Ratios   286

Form 31: Trend Analyses   288

Form 32: FFA Degrees and Participation   289

Form 33: Participation   290

Form 34: Awards and Honors   291

Form 35: The Diary   292

Form 36: Supervision of Student's Experience Program   293

**Glossary   295**

**Index   303**

# PREFACE

Financial management is important to everyone in every career. *Agribusiness: Decisions & Dollars,* Second edition is the first agricultural education high school–level text based on the Generally Accepted Accounting Principles (GAAP). The book provides students with a basis for making effective decisions, setting goals, assessing and solving problems, valuing financial progress and success, evaluating the management of resources, and gaining skills useful in everyday life.

*Agribusiness: Decisions & Dollars* follows the high school careers of two students as they pose agribusiness questions to their teacher. Their teacher, in turn, utilizes a variety of tactics to help the students understand the complexities of the financial world, teaching them that accurate records are imperative and that making effective decisions based on financial records is an important goal. The book's examples emphasize that dollar-and-sense management is a progressive need of all individuals, no matter how large or small the enterprise. The reader of the textbook learns the most basic life skills, such as the mechanics of writing a check and sequentially progress to complex enterprise analysis.

New to this edition is an introductory chapter on management and marketing that sets the stage for understanding financial management and record keeping concepts. This new chapter enables readers to broaden their appreciation of the impact of the global economy. In addition, many other edits have elevated the intensity of the content and further aligned it with appropriate dialogue and relevant examples.

The financial statements in *Agribusiness: Decisions & Dollars* were generated from the National Council for Agricultural Education's curriculum, Decisions & Dollars, and they correspond directly with the National FFA awards and degree applications, as well as with the National Management Information Systems (national agricultural education record book). My experience with each of these items (the Decisions & Dollars curriculum, the National FFA awards and degree applications, and the National Management Information System), combined with my high school and university agricultural education experience, led me to see the need for this book. The book is an excellent resource for tying together classroom work, supervised agricultural experiences, and the FFA activities.

The content for *Agribusiness: Decisions & Dollars* was pilot-tested in 10 states with almost 90 secondary-level teachers, whose input was invaluable. Based on teacher recommendations, the narrative includes the "need-to-know" information and eliminates the "nice-to-know" content. The book evolved into a one-of-kind document that can become an essential ingredient for all agricultural education programs. Its design is such that it can be used in all agricultural education classes when record keeping and financial management are addressed, or it can be the main textbook in a specific agribusiness class.

Either way, *Agribusiness: Decisions & Dollars* provides teachers with a single student reference that can simplify agribusiness education in their programs.

## CHAPTER OVERVIEW

Chapter 1    Management and marketing are the cornerstone concepts in functioning effectively in the free enterprise system.

Chapter 2    Financial management is vital in organizing and managing personal resources.

Chapter 3    Understanding the importance of inventories in the financial records of a business is critical. This chapter sets the stage for the four chapters that follow on the basic financial statements.

Chapter 4    Balance sheets, the first of the four basic statements, are an integral part of business management. A comprehensive understanding is essential as one moves through the book and learns more about other financial forms.

Chapter 5    Determining whether a business can continue to operate is a major concern. The income statement is designed to provide the profit information that is the principal ingredient in evaluating past, current, and future business performance and potential.

Chapter 6    The statement of cash flows, a relatively new financial statement, keeps a summary of cash receipts and cash expenditures, which are categorized as operating, investing, and financing activities. This analysis and planning tool is the third of the four basic financial statements.

Chapter 7    Reconciling owner equity by utilizing information from the balance sheet, income statement, and statement of cash flows provides the business with its most fundamental financial measure of its position and performance. Verifying agreement among the other three basic financial statements is the goal of owner equity.

Chapter 8    Sixteen financial measures are used to evaluate the five areas of financial analysis, liquidity, solvency, profitability, repayment capacity, and financial efficiency. Analyzing financial performance is the main reason for keeping records.

Chapter 9    Planning and decision making can be achieved in a systematic analytical process that is applicable to every phase of business and personal growth.

Chapter 10    Vast amounts of capital are needed for today's businesses. The practical and efficient use of borrowed capital is another input in the business process.

Chapter 11    Appraising the effect that income tax and other types of withholding have on wages and net income is treated as another component in total business management. The responsibility of taxes and how to prepare for them is important for all businesses.

Chapter 12    The management information system (MIS) is a combination of the financial forms found in the text and some new forms designed specifically for agricultural education students.

In addition, the accompanying workbook contains a case study, which includes a completed management information system and is a reference for educators and students while they complete their MIS in their local communities.

Agricultural education varies from school to school, from region to region, and from state to state, but there are also some commonalities. *Agribusiness: Decisions & Dollars* attempts to bring the record keeping and financial management commonalities together and introduce new concepts designed to keep agribusiness current with today's financial practices. Producing a seamless message among the classroom, supervised agricultural experiences (SAEs), and FFA requirements is a goal of this book. The hope is that this text is just the start of an evolution that brings record keeping and financial management to a heightened level of importance in the school systems.

**Jack Elliot**

# ACKNOWLEDGMENTS

The publication and success of this book are largely due to the unwavering assistance and support of many individuals. First and foremost, heartfelt gratitude goes to my wife Maureen, daughter Megan, son Nathan, daughter-in-law Katie, and granddaughter Mary Clare for their endless hours of editing and encouragement. My parents, John and Marilyn Elliot, are to be acknowledged for raising me to appreciate the value of hard work and a balance of what is really important in life. I would be remorse if I failed to thank the personnel from the National Council for Agricultural Education for providing me the opportunity to begin *DECISIONS & DOLLARS*. Finally, I wish to thank the many teachers around the country who contributed to this effort, especially the teachers from Flowing Wells High School, Stacey Rich and Curt Bertelsen, and Amphitheater High School, José Bernal and Joene Ames, in Tucson, Arizona for assisting with the many photographs used in the book.

The author and Delmar wish to thank the National Council for Agricultural Education (The Council) for permission to reprint Decisions & Dollars material for educational purposes and to incorporate Decisions & Dollars in the title of this text and supporting materials.

A thank you is also extended to the following reviewers who provided constructive comments and valuable content decisions:

Barbara Wheeling
Billings, MT

Jim Lundberg
Charles City, IA

Lauren T. Hopper
Holmer, IL

Cy Vernon
Yanceyville, NC

Joe McCain
Greenfield, IN

Larin Crossley
Preston, ID

Rebecca Carter
Rock Hill, SC

# ABOUT THE AUTHOR

**Dr. Jack Elliot** is professor and head in the Department of Agricultural Education at The University of Arizona in Tucson, Arizona. He teaches and advises students in agricultural technology management, agricultural education, and arid lands resources sciences. Jack studies the impact of high-stakes testing on academic achievement. He taught high school agricultural education in Washington and Montana. Dr. Elliot has written 34 state curriculums and one national curriculum, and he serves on local, state, national, and international educational committees. He holds a bachelor of science in agricultural education and a master of agricultural economics from Washington State University, and he is a 1988 graduate with a Ph.D. in agricultural education from Ohio State University.

# 1 Agribusiness Management and Marketing

## OBJECTIVES

*After reading this chapter, you will be able to*

- Describe the roles and functions of management.
- Identify the key economic principles of free enterprise.
- Analyze the economic opportunities of agribusiness.
- Examine the effects of government policies and regulations in making management decisions.
- Describe the management of human resources with respect to cultural diversity.
- Describe the importance of marketing.
- Explain the purpose of marketing plans.
- Develop a marketing plan.
- Discuss competitive environments and the impact of international markets.
- Compare various types of markets and their influence factors.
- Identify methods of managing risk.

## KEY TERMS

| | | |
|---|---|---|
| Demand | Free Enterprise | Market |
| Diminishing Marginal Utility | Government Policy | Marketing |
| Economic Opportunities | Hedging | Marketing Plan |
| Economic Principles | Human Resources | Supply |
| Economics | International Markets | |
| Equilibrium Price and Quantity | Management | |

# Introduction

Management and marketing are the cornerstones of any successful business. This chapter introduces some basic economic concepts and principles in agribusiness management and marketing. The dynamics of the agribusiness world are extremely complicated. The United States' agricultural industry goal is to provide a safe and reliable food supply for its citizens. Members of the agricultural industry have to be able to run their businesses by juggling finances, government policies, the influences of international markets, pest control, weather, cultural practices, and a multitude of other factors. The free enterprise system is grounded in economic principles that are the subjects of complete college textbooks and courses of study. The discussion of the concepts in this chapter provides a basic understanding of the principles and should not be interpreted as a complete or exhaustive treatment of the topics. For instance, the interaction of supply and demand is illustrated in a few simple examples, as are the concepts of diminishing marginal utility, equilibrium price and quantity, and hedging.

Economic opportunities have never been as diverse as they are in today's agribusinesses. The increase in Internet usage worldwide has placed every conceivable type of information at the fingertips of more and more agribusinesses. Understanding the influences of international agribusiness markets is made much easier with the Internet and has a direct effect on management and marketing. Successful agribusiness people utilize all of the information available as they produce marketing plans that require the flexibility to change quickly with the global influences that impact markets on a daily basis. The worldwide focus on human rights has further emphasized the importance of managing human resources at all levels of agribusiness.

This chapter outlines the characteristics of effective managers and then moves into a discussion of the functions of management including a general review of economic principles, government policies, and human resources. The second section of this chapter focuses on the marketing process and the global marketplace.

## ■ MANAGEMENT

Regardless of the definition used to describe **management,** the following terms are usually associated with it: supervising, organizing, coordinating, planning, controlling, implementing, and directing. The definition in the glossary states that management is the process of accomplishing activities associated with the goals of an agribusiness. This simple definition is expanded and detailed extensively in higher-level textbooks, but its main intent stays the same. Management is simply about the action of seeing to it that things get done.

Often in agribusinesses, the best laid plans can be destroyed by acts of nature that don't normally affect non-agribusinesses. Because of this situation and the fact that most agribusinesses deal with perishable products, the roles and functions of agribusiness managers vary slightly from managers of non-agribusinesses.

## ROLES AND FUNCTIONS OF MANAGEMENT

The agribusiness manager is the person who is responsible for the successful completion of tasks or work through people. Effective managers have the following characteristics:

1. They are technically knowledgeable about all phases of the agribusiness, especially its products or services.
2. They communicate well.
3. They motivate convincingly.
4. They are proficient in the technical skills of management, such as personnel, finance, budgeting, accounting, forecasting, and marketing.

These characteristics are necessary for managers to carry out the job of efficiently running their parts of the agribusiness. They must combine the right amounts of these characteristics for each unique situation for which they are responsible. For example, changes in government policy, the weather, or technology require the manager to adjust quickly to the new situation and make a decision on how the agribusiness will proceed. Because making good decisions is a key management role, understanding the fundamental principles that support decision making can facilitate the process.

### Making Decisions

The decision-making process is also discussed in Chapter 9. The six-step business decision-making process illustrated in Figure 1–1 outlines an effective procedure for dealing with decisions. Following the six steps does not guarantee that a decision is always the "best" one, but it does ensure that it is made in a logical and organized fashion. In fact, managers should avoid calling the selected solution the "best decision" because it is often complicated by the uncertainty of the future and the inaccuracy of the information provided at the time the decision must be made.

1. *Identifying and defining the problem.* Some problems are seasonal, such as how much to produce or what to produce. Other problems are the result of external forces, such as weather, new government policies, or technological advances. Regardless of the cause, effective managers are always cognizant of problems and potential problems. When a legitimate problem emerges, managers should focus on the problem, not on the symptoms. They need to use clear and succinct terms when defining the problem, even if they don't write it down. However, a well written problem statement enhances the entire decision-making process.

2. *Collecting appropriate information.* Some managers consider this the most important step because the ultimate decision is only as good as the analyses of the options (step 3), which relies on the input of the solid information in this step. After reflecting on the definition of the problem, the manager seeks the appropriate facts and data and assembles them into useful information. This step can be extremely time-consuming, but any attempts to curtail the collection of data and facts can lead to disastrous results. The utilization of a variety of information sources ensures that all possible avenues are used in this step. As important as this step is, managers have to establish a time line and possibly a budget to ensure that they move on to step 3.

**FIGURE 1-1**

**The six steps in making decisions.**

1. Identifying and defining the problem
   - Focus on the problem and not the symptoms
   - Use clear and succinct terms when defining the problem
2. Collecting appropriate information
   - Gather data and facts
   - Refer to the concise definition of the problem during this step
3. Identifying and analyzing alternative solutions
   - Prioritize measurable criterion (e.g., units produced, sales, costs, profits, etc.) that will be used to evaluate the alternative solutions
   - Utilize the data and facts to assemble a list of possible solutions
4. Selecting an alternative
   - Based on the selected measurable criterion and the collection of appropriate data and facts, an informed decision can now be made
5. Implementing the decision
   - Obtain the associated resources to carry out the new decision
   - Establish a process to ensure that the solution is implemented in a timely manner
6. Evaluating the decision
   - Refer to the original definition of the problem
   - Determine if the problem was effectively addressed
   - Assess steps that would have enhanced the solution
   - Improve the manager's decision-making abilities for the next problem

3. *Identifying and analyzing alternative solutions.* Brainstorming all possible ideas after reviewing the compiled information from step 2 should lead to a list of solutions (Figure 1–2). Managers must prioritize measurable criteria (e.g., units produced, sales, costs, profits, etc.) that will be used to evaluate the alternative solutions. They need to analyze the information and not be restricted by traditions or habits.

4. *Selecting an alternative.* An informed decision is just that: a decision based on the information that the manager has at one point in time. The decision not to do anything is a decision, as is the decision to go back and solicit more information. Often, managers are faced with several options that show promise but that rely on the uncertainty of the future. In this case, managers must assess the risks involved in selecting one potential outcome over another.

5. *Implementing the decision.* Efficiency and timeliness are the two important factors in step 5. If the first four steps are completed in an effective and comprehensive manner, then a road map for implementation should be easy to develop. For instance, if during step 2 it was determined that an influx of capital funds is necessary to implement the solution, the manager has only to execute that task. Likewise, all of the important operations to successfully achieve the new goals of this decision should be outlined in steps 3 and 4.

**FIGURE 1-2**

These students use their managerial skills to repair the greenhouse cooling system.

6. *Evaluating the decision.* Managers are responsible for making decisions. Therefore, it is in their best interest to take considerable time in assessing how well the decision is implemented. Did implementation accomplish the task addressed in the original problem statement? Are there more efficient and effective ways to deal with the problem? What can be done to improve the manager's decision-making abilities for the next problem? A careful evaluation will result in more current and accurate information that can be utilized the next time a decision must be made.

Decisions can generally be classified as either organizational or operational. Developing business plans and procuring resources are types of organizational decision. Examples include how big a loan to take out, what types of crops or livestock to raise, and how much land to purchase or lease. Usually, these types of decisions are made on an annual basis.

Operational decisions are made much more frequently and are essential in the day-to-day running of the agribusiness. Examples include establishing employee work schedules; selecting dates for planting, harvesting, fertilizing, and other tasks; marketing products; and making equipment repair decisions (Figure 1–3).

Other decision-making factors include how important the decision is, how frequently the problem occurs, how quickly the decision has to be made, and whether the decision can be changed once it is made. By understanding all these factors, managers can make effective and efficient decisions that are in the best interest of the agribusiness.

**FIGURE 1-3**

The decision to plant these flowers in September ensured their readiness for the winter holiday market.

## Functions of Management

The functions of management correspond directly to the decision-making process. Figure 1–4 outlines the three main functions of management:

1. planning and organizing
2. deciding and implementing
3. analyzing and evaluating

The first function involves planning and organizing all endeavors associated with the agribusiness. This function is often concerned with forecasting and what will happen in the future, but given the many problems that can occur in

**FIGURE 1-4**

Functions of management.

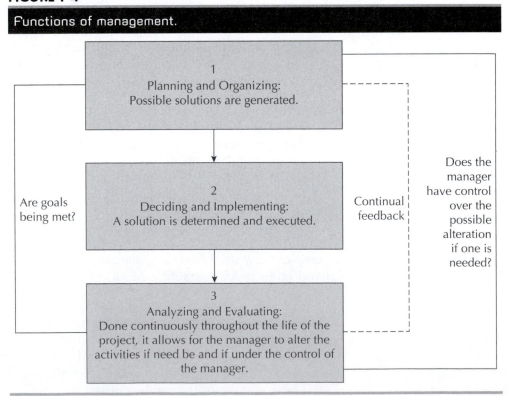

agribusinesses, implementing the decision-making process to find quick solutions is also a common activity among managers. Managers are responsible for any number of activities and problems that seem to arise at the same time. Planning and organizing in order to forecast requires a solid understanding of the agribusiness, including production, inventory, marketing, and financial management. The organizational skills utilized by managers vary according to the size of the agribusiness and the complexity of the activity or existing problem. For example, in a mega-agribusiness there may be many departments and many layers of management, whereas in a small family-owned agribusiness the decision may affect only one or two employees. Regardless of the size of the firm or the type of activity or problem to be addressed, the goal of the first function is to generate possible solutions.

Once the solutions are generated, the second managerial function is employed: deciding and implementing the selected solution (Figure 1–5). In large agribusinesses this could mean acquiring additional resources, reorganizing entire departments, and hiring more employees. In small family-owned agribusinesses, implementing a new solution may require minimal adjustments in daily operations. The primary goal of the second managerial function is to select a solution and attempt to make it work.

The final function of analyzing and evaluating is called the control function in many textbooks. An accurate and up-to-date management information system must be in place so that managers have the detailed information required to determine whether the solution is being effectively implemented. Continuous monitoring may result in a modification of the original plan, maybe several times. A question that managers must ask during this phase is, "Do I have control over the suggested alteration?" In many cases, the agribusiness may have to live with the current situation because the proposed change is out of its control, not feasible or timely, or cost prohibitive.

Because of the complexities facing today's agribusinesses, a great deal of attention is devoted to managing a business. Understanding the decision-making

**FIGURE 1–5**

The decision to adjust ventilation is a management decision.

process and the roles and functions of management are important steps for successful agribusiness managers. Operating agribusinesses in the United States requires a basic appreciation of **economics**.

## KEY ECONOMIC PRINCIPLES OF FREE ENTERPRISE

Demand for goods and services is the cornerstone of the American **free enterprise** system. Effective managers are able to identify the needs of consumers and adjust their agribusinesses to meet those needs quickly and efficiently. A basic **economic principle** implies that consumers seek satisfaction or utility from all they consume. They select products or services that provide high levels of satisfaction. Another basic economic principle suggests that the amount of satisfaction diminishes as consumption of additional units of a product or service continues. For example, the first bite of a pizza provides great satisfaction, but each additional bite provides less and less satisfaction until consumers can't possibly eat another bite. In fact, consumers might reach a point where they would pay not to eat another bite. In other words, the marginal utility of consuming each bite after the first bite continues to lessen. The phenomenon of diminished satisfaction with each bite of pizza is called **diminishing marginal utility.**

In the free enterprise system, price and quantity adjust themselves to consumer **demand.** In Figure 1–6 the demand schedule illustrates that the higher the price goes, the lower is the quantity demanded. Inversely, the lower the price is, the higher is the quantity demanded. In this example, a price of $10 results in a demand of 10 units, whereas a price of $2 yields a demand of 50 units. Several items can influence demand, such as whether there is a replacement or complement product, whether the population is increasing or decreasing, whether income is improving or changing, and whether timing affects demand (e.g., selling holiday items is effective only immediately prior to the event).

A shift in the demand schedule (Figure 1–7) from Q1 (Quantity 1) to Q2 (Quantity 2) is usually attributed to changes in the items that influence

**FIGURE 1-6**

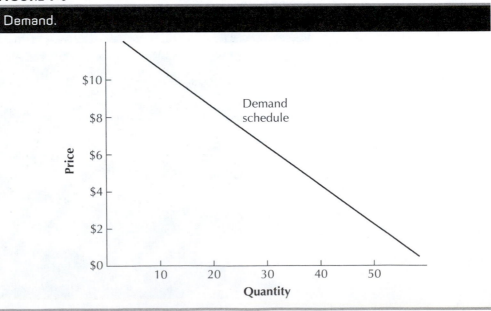

Demand.

**FIGURE 1–7**

Demand shift.

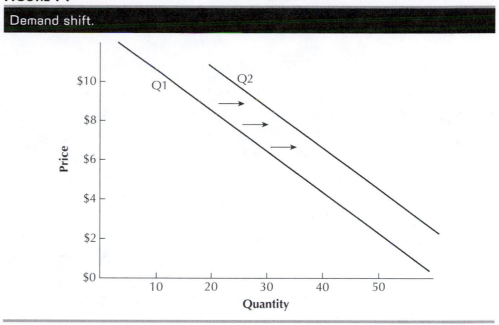

demand (mentioned in the previous paragraph). For instance, suppose a product is selling at $6 and 30 are sold, then suddenly 40 units are being sold. The response is attributed not to price, but to one or more of the influential factors or demand shifters. In this example, let's say that consumers in an area just received a cost of living increase, which may result in more people eating out during the following month. The increase or shift in demand is a result of more income, not of any price changes by the restaurants.

The demand schedule is derived from consumer actions, and the **supply** schedule is derived from producer actions. As shown in Figure 1–8, as price increases, producers are apt to generate more products or services. Likewise,

**FIGURE 1–8**

Supply.

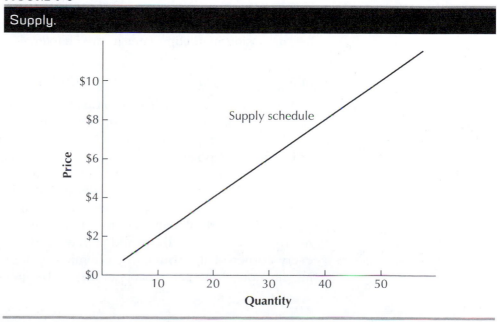

**FIGURE 1-9**

Interaction of demand and supply.

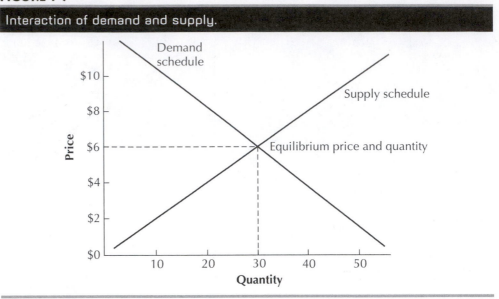

as price decreases, producers are apt to produce more quantity. The supply schedule can be overlaid on the demand schedule to create an interaction of supply and demand and to predict the **equilibrium price and quantity** (Figure 1–9). This phenomenon theoretically occurs when the market exists without hindrance from outside forces and when the demand and supply schedules intersect.

In theory, at the equilibrium price, the quantity offered for sale is exactly equal to the quantity buyers are willing to buy. If the price increases, buyers simply do not purchase the entire supply available. If the price decreases, producers do not have enough product, and a shortage results. In reality, changes in supply and demand occur continuously for many reasons, and effective managers are vigilant in about staying up-to-date on the latest economic trends. This very simple introduction to the key economic principles provides a basic understanding of the free enterprise system, as well as a foundation for discussing marketing later in the chapter.

## Economic Opportunities of Agribusiness

Agribusiness **economic opportunities** have never been greater. The global economy erases borders for many goods and services. Everyone in the world needs to eat, and providing safe and nutritious products is a challenge because of cultural and political differences. For example, the debate over genetically modified products influences markets and needs to be addressed by managers during the business planning phase. Seeking niche markets with specialty items can provide the innovative manager with economic opportunities never imagined in previous generations. Marketing and information seeking via the Internet have brought the world markets and relevant data and facts to almost every corner of the country. Economic opportunities are limited only by the imagination of the participants in the agribusiness.

## Government Policies and Regulations

The United States Department of Agriculture (USDA) is the primary agency that provides economic **government policies** and regulations for agribusinesses. Agribusiness managers must pay close attention to the local USDA policies and regulations because these requirements can have huge economic effects on their businesses. In many cases, there may be the difference between staying in business and being forced out of business. Because the policies and regulations are ever changing, and because there are vast regional differences among them, managers have to be aware that they exist and keep up-to-date on the ones that affect them.

## Human Resources

Sixteen percent of the gross national product can be attributed to the agriculture and the food industry, and it takes people to make agribusinesses run. Although most of this chapter has focused on managers, most of the people involved in agribusiness are not managers, but workers of some type. The importance of having productive **human resources** in agribusinesses cannot be understated. It is people who operate and repair equipment. It is people who plant and harvest crops. It is people who care for livestock. It is a very diverse population of people who make American agribusinesses among the most productive in the world. Being a manager in an agribusiness means understanding the cultural diversity of the people in the industry. Some cultures insist on not working on Sunday. Other cultures focus heavily on family activities. An effective manager must be sensitive to cultural activities and plan accordingly.

## ■ MARKETING

The definition of **marketing** is the coordination of agricultural production with consumer demand including the conception, pricing, promotion, and distribution of products and services. Predicting consumer demand is the key element in marketing (Figure 1–10). Entire companies are devoted to this very concern. The many steps in the marketing process go from research to production to consumer consumption. Assembling, processing, wholesaling, and retailing are all possible economic phases of the marketing system. Marketing is very complex, and this introduction provides the very basics needed to create a marketing plan.

## IMPORTANCE OF MARKETING

Essentially, satisfying consumer needs drives the American economy and illustrates the importance of marketing. Most businesses adopt a management process that answers the question, "Are we satisfying consumer needs?" The answers to this question impact management decisions throughout the entire company. To elaborate on the question, managers want to know how the consumer wants a product prepared, where the consumer can find it, when the consumer wants it, and what is the easiest way for the consumer to obtain it (Figure 1–11). These marketing questions, or stages in the marketing system, are called utilities.

**FIGURE 1–10**

Satisfying customer demand for fresh vegetables is made easier with climate-controlled production.

**FIGURE 1–11**

Marketing questions or utilities.

- Form utility: how the consumer wants a product prepared?
- Place utility: where the consumer can find it?
- Time utility: when the consumer wants it?
- Possession utility: what is the easiest way for the consumer to obtain it?

## MARKETING PLANS

Functional **marketing plans** enable agribusinesses to successfully market their goods or services. Figure 1–12 outlines the three market plan components. The first step is to develop a purpose statement that clearly identifies what consumer demand or need the agribusiness will meet. The second step is to create a set of objectives that concisely states how the agribusiness will meet the consumer demand or need. The uniqueness of the agribusiness can be illustrated in its objectives, such as offering the best service, the largest selection, lowest prices, and the like. This step helps establish a competitive edge for the agribusiness.

The third marketing plan step is the most comprehensive and may involve hiring market research companies to attain some of the answers. Understanding all facets of the potential market is a crucial component of the plan. For instance, determining the potential customers—what they want and when they want it—is the first question that needs to be answered in this step. A marketing area must be determined, and the competition identified and understood. The more attention that is directed toward obtaining accurate and complete answers to these questions, the better the marketing plan will be (Figure 1–13).

**FIGURE 1-12**

Marketing plan.

1. What consumer demand or need will the agribusiness devote its energies and resources toward?
   a. The firm's purpose statement should clearly and briefly state how it will address the question.
   b. An example would be, "The purpose of Sweeney Insurance is to provide producers with crop insurance."
2. How will the agribusiness meet the identified consumer demand or need?
   a. Objective statements must also be clear and brief, and illustrate exactly how the consumer needs will be met.
   b. For instance, "Sweeney Insurance will offer competitive prices and efficient service for crop insurance."
3. Identify the potential market:
   a. Who are the potential customers?
   b. What do they need?
   c. When do they need it?
   d. How big is the area served?
   e. Who is the competition?
   f. What does it cost to operate the business?
   g. What does it cost the competition to stay in business?
   h. What might your competitors do in response to your marketing strategies?

**FIGURE 1-13**

Once a marketing plan is written, advertising becomes an essential part of informing consumers of your products.

## COMPETITIVE ENVIRONMENTS AND INTERNATIONAL MARKETS

**International markets** can provide additional outlets for U.S. agribusiness products and services, and they also compete on American soil for a piece of the domestic market. For many small and local agribusinesses, the international influence is immaterial to their operations. However, large agribusinesses' understanding of the impact of international markets can determine the difference between success and failure.

Understanding international markets requires studying their economic, sociocultural, political-legal, technological, and competitive environments. Each of these environments is extremely complex, and their complete comprehension often means living abroad and immersion in the international communities. Regardless of the product or service, the global economy impacts domestic prices, quality, and distribution, and it is a part of doing business in today's world. International markets are neither good nor bad. They are only an additional factor in the operation of businesses in the free enterprise system. Acknowledging their existence and understanding the complexities associated with their impact are major steps in the creation of a successful agribusiness.

## TYPES OF MARKETS

There is an almost unlimited number of agribusiness **markets** in the world. The following six market environments reflect an attempt to categorize them:

1. *Open markets* include commodity futures markets, auctions, stock exchanges, and the like. In an open market, prices are based on the interaction of buyers and sellers at a specific point in time. Many factors influence the final price of the product or service, including the amount of competition, the amount and quality of the supply, and the urgency of the demand.

2. *Formula pricing markets* utilize a formula to generate the price. Often, a public or partially public group establishes control in this type of market. The state and federal milk marketing groups and retail milk commissions are examples.

3. *Legislated markets* are commodity markets, in which the government mandates either or both minimum and maximum prices by means of fair trade laws, price support programs, or minimum wage laws. Some commodities that are traded in legislated markets are wheat, soybeans, peanuts, cotton, rice, and feed grains. The legislated prices do not necessarily follow or support a free enterprise–generated price. That is, they are sometimes higher and sometimes lower than the price would be if it were left to the market for determination.

4. In *group disciplined markets*, professional or occupational organizations establish prices for their members, usually charge for services, and impose penalties if members deviate from deadlines or other group rules. Legal and medical societies, as well as some unions, are included in this type of market environment.

5. *Negotiated markets* achieve price levels through formal negotiations between buyers and sellers. A very formal example is collective bargaining between labor unions and management. Another less formal example is an agreement between producers and agribusinesses that sell the product to the public. Some commodities that are bought and sold in this type of market are poultry, eggs, vegetables, fruit, and some natural resource products.

6. *Regulated markets* are very important to the American economy and include many public services such as water, electricity, bus and railroad fares, natural gas, or the price of freight. These prices are stipulated by the Interstate Commerce Commission and state public service commissions.

## MANAGING RISK

Perhaps the most difficult marketing function is managing risk. Several types of risks affect agribusinesses, such as protecting assets from physical damage or loss. For example, agricultural producers insure their crops in various ways to protect them from weather or pest damage. For this book, the risk focus is on price risk.

Price risk is a big issue in agribusinesses. Unless managers operate in the futures markets, where the risk can be transferred to others, they are destined to accept the current market price, which can fluctuate depending on the season. For instance, producers who sell their grain during harvest, when there is a glut or a high supply of grain on the market, can expect much lower prices than later in the year when the demand for grain is still high, but the harvest is over and stored grain can command a higher price.

Using the futures market through hedging or futures trading is another way to share price risks. Prior to the harvest, producers can sign a futures contract for a guaranteed price with a harvest delivery. The example in Figure 1–14 illustrates that **hedging** protects producers from great losses but hinders them from benefiting from huge profits if the price goes up. If the price skyrockets, there are ways to buy back the futures contract and capture some additional profits, but that concept is for a higher-level textbook. The point is that

**FIGURE 1–14**

### Futures markets or hedging.

The following scenarios illustrate the potential gross income from the same amount of wheat.

- During planting in April the producer sells 20,000 bushels of wheat for August delivery at $3.50 per bushel.    $70,000 gross income
- During August the producer sells 20,000 bushels of wheat for the cash market price of $2.50 per bushel.    $50,000 gross income
- During August the producer sells 20,000 bushels of wheat for the cash market price of $4.50 per bushel.    $90,000 gross income

usually crop prices are depressed during harvest because of the high supply. Therefore, producers who hedge their crops by selling them on the futures market usually receive higher prices and minimize the price risk of growing and marketing the crop.

## Summary

Management can be categorized as a process of three interrelated functions: planning and organizing, deciding and implementing, and analyzing and evaluating. Managers employee decision-making skills as they make plans for the future and solve daily problems. Having a solid understanding of the interactions of supply and demand enhances managers' abilities to effectively function. Influences other than economic factors, such as government policies and regulations, also influence the decisions managers make. Managing human resources remains an important function in successful agribusinesses.

The importance of marketing and its relationship to management is directly tied to consumer demand. Agribusinesses that focus on satisfying consumer demand are usually successful in the free enterprise system. Coordinating consumer demand with the production of goods and services defines marketing. The ultimate goal of marketing is high efficiency and consumer happiness.

This introductory chapter introduces basic management and marketing concepts and states that having a sound financial management background is essential for successful agribusinesses. The remainder of the book is devoted to financial management. In Chapter 2, Troy and Megan are introduced as two students of Ms. Green. Their conversation continues through the rest of the book, and their questions and observations are similar to ones that you might have as you learn sound financial management principles.

### SUPERVISED EXPERIENCE

Supervised experiences in occupationally driven agricultural education programs allow students to apply classroom content in a realistic situation. These programs are focused on individual or group interest, and they are supervised by an instructor, employer, and/or parent/guardian. Students who participate in supervised experiences keep many types of records, such as financial, production, time-invested, and diary.

### CAREER OVERVIEW

Most agriscience careers require a sound understanding of financial management because agribusinesses, like any business, have to compete in the marketplace. Some agriscience careers focus almost entirely in the financial arena, some careers are centered around production agriculture, and some careers are almost solely scientific. The careers listed in this book illustrate opportunities that require financial, scientific, and production agriculture backgrounds.

High school diploma: Fertilizer Plant Worker; Agricultural Machinery and Equipment Operator

Associate's degree: Bookkeeping Specialist; Agricultural Sales

Bachelor's degree: Public Relations Director; Agricultural Lobbyist

Advanced degree: International Relations; Commodity Director

## DISCUSSION QUESTIONS

1. List the four characteristics of effective managers.
2. Identify and explain the six steps in the decision-making process.
3. Provide examples that delineate between organizational and operational decisions.
4. Describe the three managerial functions.
5. Illustrate a simple interaction of demand and supply.
6. Identify and explain the four marketing questions or utilities.
7. List the six market types.

## SUGGESTED ACTIVITIES

1. Examine the local effects of government policies and regulations in making agribusiness management decisions. Ask local agribusiness managers, government personnel, and financial institution employees for their views on the subject, and compare their responses. This may be done as a class project.
2. Analyze the management of human resources with respect to the cultural diversity of local, state, or regional agribusinesses.
3. Develop a marketing plan for your supervised experience program, a school enterprise, or a local agribusiness.
4. Interview agribusiness managers, and determine the status of hedging in your local community.
5. Determine which of the six market types is the most prominent in your community as it relates to your local agribusiness economy.

## GLOSSARY

**Demand** The relationship between price and quantity that illustrates how much consumers are willing and able to purchase at various prices.

**Diminishing Marginal Utility** When a product or service is consumed, a time comes when the utility (or pleasure or benefit) of consuming that product or service declines.

**Economic Opportunities** The possible business prospects available in today's global economy.

**Economic Principles** The basic premises that explain the free enterprise system.

**Economics** The science of allocating and utilizing scarce resources of land, labor, capital, and management in the most optimum manner among different and competing choices to satisfy human wants.

**Equilibrium Price and Quantity** The point at which the supply and demand schedules intersect and at which the quantity offered for sale matches exactly the quantity that consumers are willing to purchase.

**Free Enterprise**   A system in which economic agents are free to own property and engage in commercial transactions.

**Government Policy**   United States Department of Agriculture policies that are designed to regulate agribusinesses.

**Hedging**   To offset risk in the market, hedgers use the forward or futures market to guarantee a transaction or price and thereby reduce exchange risk.

**Human Resources**   The people who work in agribusinesses. Labor used as an input into the production process.

**International Markets**   All markets outside the borders of the United States.

**Management**   The process of accomplishing activities associated with the goals of the agribusiness.

**Market**   The interaction between supply and demand to determine the market price and corresponding quantity bought and sold.

**Marketing**   The coordination of agricultural production with consumer demand including the conception, pricing, promotion, and distribution of products and services.

**Marketing Plan**   An agribusiness's written plan of action that details the time frames and activities for implementing a strategy to market its goods and/or services.

**Supply**   The relationship between price and quantity that illustrates how much producers are willing to supply at various prices.

# 2

# Developing Personal Life Skills

## OBJECTIVES

*After reading this chapter, you will be able to:*

- Categorize personal income and expenses.
- Prepare a personal budget to accomplish given financial objectives.
- Accurately write a check and record the transaction in a checkbook ledger.
- Accurately complete a deposit slip and a savings withdrawal slip.
- Reconcile a bank statement and checkbook ledger.
- List the reasons for savings and investments.
- Compare characteristics of various types of investments.
- Explain the concept of interest.
- Describe how investments can grow in value and calculate the future value of money.
- Establish a personal goal for savings and investments.
- Discuss the importance of the ability to obtain credit.
- Distinguish between types of credit.
- Identify possible sources of credit.
- Establish good credit.
- Differentiate between secured and unsecured loans.
- Compare interest rates of various personal loans.
- Identify employer expectations, appropriate work habits, and good citizenship.
- Describe personal and occupational safety practices in the workplace.

# BUSINESS PROFILE

*J*anet Murphy's career goal is to be a veterinarian. After enrolling in agriscience, her teacher set up a Supervised Agricultural Experience (SAE) program for her. The experience consists of working as a veterinarian assistant in her town after school and for a half day on Saturday. The veterinarian agreed to pay her $4.75 per hour and to teach her about caring for animals. When she began the job she opened a checking account. She shopped around and found the lowest rate ($5.25 per month) for maintaining a checking account at the bank. Janet received her own personalized checks with her account and the cost ($10.83) was to be deducted from the account. Also, she received a card to use at an automated teller machine (ATM), and she withdrew $10.00 to test the card. In class Janet learned the importance of keeping good records and was determined to keep an accurate accounting of her finances. In her checkbook she listed every check she wrote. Before the end of the month, she was surprised to get a letter from the bank telling her that she was overdrawn. Because she had kept a record of all the checks she wrote, she could not figure out why the bank would think she had overdrawn. Following are the records in her checkbook. Can you figure out the problem?

| Date | Description | Check Amount | Withdrawal | Deposit | Balance |
|------|-------------|--------------|------------|---------|---------|
| April 3 | Paycheck | | | $183.68 | $183.68 |
| April 5 | John's Pizza Parlor | $11.26 | | | $172.42 |
| April 8 | Jill's Dress Shop | $83.37 | | | $ 99.05 |
| April 18 | Acme Pet Store | $13.77 | | | $ 85.28 |
| April 20 | Year Book | $28.98 | | | $ 56.30 |
| April 20 | Banner Book Store | $14.88 | | | $ 41.42 |
| April 21 | Gala Music Shop | $30.88 | | | $ 10.54 |

**SOLUTION:** Janet made a $10.00 calculating error when she subtracted the $83.37 written to Jill's Dress Shop. The correct balance was $89.05, not $99.05. Also, she forgot to subtract the $10.83 for the cost of her checks and the $5.25 for the monthly service charge. Also, she forgot to record the $10.00 withdrawal from the ATM. She should double-check her figures and also keep track of all expenses associated with the account.

| Corrected Check Register | | | | | |
|---|---|---|---|---|---|
| **Date** | **Description** | **Check Amount** | **Withdrawal** | **Deposit** | **Balance** |
| April 3 | Paycheck | | | $183.68 | $183.68 |
| April 5 | John's Pizza Parlor | $ 11.26 | | | $172.42 |
| April 8 | Jill's Dress Shop | $83.37 | | | $ 89.05 |
| April 18 | Acme Pet Store | $13.77 | | | $ 75.28 |
| April 20 | Year Book | $28.98 | | | $ 46.30 |
| April 20 | Banner Book Store | $14.88 | | | $ 31.42 |
| April 20 | ATM withdrawal | | $10.00 | | $ 21.42 |
| April 21 | Gala Music Shop | $30.88 | | | ($ 9.46) |
| | Cost of checks | | $10.83 | | ($ 20.29) |
| | Monthly service charge | | $ 5.25 | | ($ 25.54) |

## KEY TERMS

| | | |
|---|---|---|
| Annual Percentage Rate (APR) | Disbursement | Net Worth |
| Annuity | Discounting | Present Value |
| Balance Sheet | Dividend | Principal |
| Bond | Future Value | Profit |
| Certificate of Deposit | Interest | Risk |
| Check Register | Investment | Savings Account |
| Checking Account Statement | Liquidity | Stock |
| Collateral | Market Value | Tax |
| Compounding | Maturity | Time Value of Money |
| Credit | Money Market Account | Treasury Bill |
| Credit Union | Municipal Bond | Treasury Bond |
| Direct Deposit | Mutual Fund | Treasury Note |

## Introduction

The purpose of this chapter is to introduce personal financial management. Financial management is vital in organizing and managing your personal resources. The chapter emphasizes the significance of personal financial management and why it should be a priority in one's everyday life (Figure 2–1). For example, you will learn how to open and properly manage a checking and/or **savings account.**

Another purpose of this chapter is to study some of the common alternatives when making personal **investments.** Investments come in various forms. Investments such as **money market accounts,** savings accounts, **certificates of deposit,** and savings bonds are common personal alternatives. The stock market is another form of investment. The most important factor in investing is that the investment must meet the requirements of the investor (i.e., **profit,** risk, security, availability, and the like).

Understanding the reasons for obtaining **credit** follows logically after the section on investing. Credit cards have become a natural part of our everyday lives. The personal loans available through credit cards are convenient for individuals, but they have some major drawbacks. Personal credit provides "wheels" as individuals deal with financial tasks and find themselves in motion in the nonstop monetary world. When credit is appropriate and necessary, knowing the criteria required to obtain personal loans and credit cards becomes important. To obtain credit, your creditor must consider you an acceptable financial risk.

An important part of developing personal life skills is preparing for the working world. Employers have definite expectations, and one should follow appropriate work habits. In addition, following personal and occupational safety practices is critical to being successful as an employee.

**FIGURE 2–1**

An important purpose of this book is to provide students with the abilities to make personal and business financial decisions. Evaluating feed costs is essential when operating a livestock enterprise.

## ■ PERSONAL FINANCIAL MANAGEMENT

### REASONS FOR PERSONAL FINANCIAL MANAGEMENT

The primary reason for keeping personal financial records is to acquire information that is needed for management decisions. Knowing one's daily financial standing is imperative when setting financial priorities and needs (see Tables 2–1 and 2–2). In addition, it is important for sound credit planning. For example, having an accurate idea of cash flow is essential to identify a business's credit needs and repayment capacity. Charting financial progress with the 16 financial ratios from Chapter 8 is made possible by accurate financial record keeping. Accurate income and depreciation records make **tax** reporting easy. Furthermore, it allows for easy verification of the truthfulness of tax reports in the case of an audit. Two students, Troy and Megan, and their teacher Ms. Green will be with you throughout the book. The student's questions will be addressed by Ms. Green or by her colleagues and friends.

Ms. Green teaches suburban high school students, who come from both the city and the surrounding rural area. She offers a challenging agriscience curriculum that embraces the latest scientific principles in an applied academic setting and that utilizes plenty of hands-on applications. Troy, from the city, and Megan, from the rural area, represent the diversity of the students in her program.

### TABLE 2-1   WHERE DOES YOUR MONEY COME FROM?*

**Troy: first week of school**

| Sources of Income | Percent of Total Income |
| --- | --- |
| Parent—allowance.............................$35.00 | 100% |
|  |  |
|  |  |
|  |  |
|  |  |
|  |  |
|  |  |
|  |  |

*Source: The National Council for Agricultural Education, DECISIONS & DOLLARS, 1995

### TABLE 2-2   WHERE DOES YOUR MONEY GO?*

**Troy: first week of school**

| Item Purchased | "Luxury" or "Required" |
| --- | --- |
| Baseball cards.....................................$10.00 | Luxury |
| Lunch (5 days @ $2.00)......................$10.00 | Required |
| Junk food and snacks..............................$15.00 | Luxury |
|  |  |
|  |  |
|  |  |
|  |  |
|  |  |
|  |  |
|  |  |
| Total................................................$35.00 | Luxury $25.00................... Required $10.00 |

*Source: The National Council for Agricultural Education, DECISIONS & DOLLARS, 1995

Troy is an ambitious young man who dreams of operating his own horticulture business. Living in the city provides him work experience opportunities and offers him a market for his plants, but he lives with his parents in an apartment and has no room for a greenhouse.

Megan lives with her family on a small acreage outside the city limits. She is unsure of what she wants to do with her life, but she heard that Ms. Green is a good teacher and decided to join some of her friends and enroll in a beginning agriscience class.

*T*roy raised his hand and asked, "Why is it important to keep track of money? I seem to have enough money for everything I want."

"*T*hat may be true now, but very shortly you will be on your own or you will want to purchase something that will require a savings account. In addition, many good problem-solving skills can be learned through keeping records and making management decisions. Let's see where your money came from this week and where it went by completing Tables 2–1 and 2–2," Ms. Green directed.

## ADVANTAGES AND DISADVANTAGES OF PERSONAL FINANCIAL MANAGEMENT

Accurate assessment of personal financial standing, or **net worth,** is easier to calculate when records are kept. Information for sound financial planning and budgeting is also available from accurate records. However, it takes time to set up a system. Time must be scheduled on a regular (monthly) basis for effective record keeping. Decisions must be made as to whether a person or business utilizes one of the various computer "personal financial management" systems, that is, programs for financial record keeping, check writing, and decision making (Figure 2–2). Non-computerized accounting systems still exist and are quite adequate for record keeping in many instances.

**FIGURE 2-2**

Setting up a computer-based financial record-keeping system takes time, but it improves efficiency and accuracy.

"**I** spent $25 on luxury items in one week. If I saved that much each week I could have $100 in only one month," Troy exclaimed.

**M**s. Green responded, "That's right, sometimes if you keep track of actual financial transactions the figures can be quite revealing. A better way to keep track of financial matters is by recording revenue or income and costs or expenses on a ledger like Table 2–3. Let's look at Megan's ledger (Table 2–4). She didn't receive a big allowance like you, but she was still able to save a little money during her first two weeks of school. Let's enter your financial transactions in Table 2–5 and calculate your balance."

**M**egan observed, "I saved a quarter of my allowance by watching what I spend."

**TABLE 2-3   PERSONAL FINANCIAL MANAGEMENT FORM***

| Date | Item | Income | Expenses | Balance |
|------|------|--------|----------|---------|
| | Beginning Balance | | | |
| | | | | |
| | | | | |
| | | | | |
| | | | | |
| | | | | |
| | | | | |
| | | | | |
| | | | | |
| | | | | |
| | | | | |
| | | | | |
| | | | | |
| | | | | |
| | | | | |
| | | | | |
| | | | | |
| | Ending Balance | | | |

*Source: The National Council for Agricultural Education, DECISIONS & DOLLARS, 1995

**TABLE 2–4    PERSONAL FINANCIAL MANAGEMENT FORM (MEGAN)\***

**Megan: first two weeks of school**

| Date | Item | Income | Expenses | Balance |
|------|------|--------|----------|---------|
| Sept. 5 | Beginning Balance | | | $25.00 |
| 12 | Allowance | 5.00 | | 30.00 |
| 12 | Lunch | | 2.00 | 28.00 |
| 13 | Lunch | | 2.00 | 26.00 |
| 15 | Lawn mowing—Smiths | 10.00 | | 36.00 |
| 15 | Movie/snacks | | 9.50 | 26.50 |
| 16 | Lunch | | 2.00 | 24.50 |
| 17 | Lunch | | 2.00 | 22.50 |
| 19 | Allowance | 5.00 | | 27.50 |
| | | | | |
| | | | | |
| | | | | |
| | | | | |
| | | | | |
| Sept. 19 | | | | |
| | Ending Balance | $20.00 | $17.50 | $27.50 |

\*Source: The National Council for Agricultural Education, DECISIONS & DOLLARS, 1995

## ■ CHECKING ACCOUNTS

The ledger formats from Tables 2–3, 2–4, and 2–5 provide an up-to-date balance, much like a **checking account statement** (Table 2–6). Checking accounts offer several convenient features for handling personal finances. They are

■ an alternative method of payment other than cash.

■ a convenient and safe method for payment by mail.

■ an acceptable form of payment at most businesses, provided appropriate ID is presented (e.g., a valid driver's license).

■ a means of listing **disbursements** and deposits.

■ sometimes a source of **interest** on account funds.

■ sometimes accompanied by "special incentive" for opening the account.

A variety of types of checking accounts are available to meet your personal and business needs. There are

■ no-fee small personal accounts.

■ interest-earning accounts.

**TABLE 2-5    PERSONAL FINANCIAL MANAGEMENT FORM (TROY)***

Troy: first two weeks of school

| Date | Item | Income | Expenses | Balance |
|---|---|---|---|---|
| Sept. 5 | Beginning Balance | | | $80.00 |
| 12 | Allowance | 35.00 | | 35.00 |
| 12 | Baseball cards | | 10.00 | 25.00 |
| 13 | Lunch | | 2.00 | 23.00 |
| 14 | Lunch | | 2.00 | 21.00 |
| 15 | Lunch | | 2.00 | 19.00 |
| 16 | Lunch | | 2.00 | 17.00 |
| 17 | Lunch | | 2.00 | 15.00 |
| 18 | Junk food and snacks | | 15.00 | 0.00 |
| | | | | |
| | | | | |
| | | | | |
| | | | | |
| | | | | |
| | | | | |
| Sept. 19 | Ending Balance | $35.00 | $35.00 | $80.00 |

*Source: The National Council for Agricultural Education, DECISIONS & DOLLARS, 1995

■ package accounts that combine services such as checking, savings, money markets, individual retirement accounts (IRAs), earned interest, ATM use, overdraft protection, a check guarantee card, and so on (Figure 2–3).

There are a variety of check and checkbook styles.

■ *Duplicate.* A carbon copy of the check remains in the checkbook.
■ *End/top stub.* The stub, with check information, remains in the checkbook.
■ *Safety paper.* The checks are printed on watermark paper (a photocopy would be obvious).
■ *Desk set.* The checks come three to a page in a large desk-sized binder.

"*I*'m sure I don't have enough money to open a checking account," commented Megan. Ms. Green answered, "Financial institutions, such as banks, often offer young people special, low, or zero balance checking accounts. This is done to help students become familiar with the bank and the services it provides."

## TABLE 2-6   CHECKBOOK LEDGER*

Troy: first two weeks of school

| | | | | | | | | Balance | |
|---|---|---|---|---|---|---|---|---|---|
| **PLEASE BE SURE TO *DEDUCT* ANY PER CHECK CHARGES OR SERVICE CHARGES THAT MAY APPLY TO YOUR ACCOUNT** | | | | | | | | | |
| **Number** | **Date** | **Checks Issued to or Description of Deposit** | **(−) Amount of Check** | | **✓ T** | **(−) Check Fee (if any)** | **(+) Amount of Deposit** | **0** | **00** |
| | 9/12 | Allowance | | | | | 35  00 | 35 | 00 |
| | | | | | | | | 35 | 00 |
| 00001 | 9/12 | Tommy Smith | 10 | 00 | | | | −10 | 00 |
| | | Baseball cards | | | | | | 25 | 00 |
| 00002 | 9/13 | Cash | 10 | 00 | | | | −10 | 00 |
| | | For school lunch | | | | | | 15 | 00 |
| 00003 | 9/18 | Cash | 15 | 00 | | | | −15 | 00 |
| | | For junk food, snacks, and fun | | | | | | 0 | 00 |
| | | | | | | | | | |
| | | | | | | | | | |
| | | | | | | | | | |
| | | | | | | | | | |
| **REMEMBER TO RECORD AUTOMATIC PAYMENTS/DEPOSITS ON DATE AUTHORIZED.** | | | | | | | | | |

*Source: The National Council for Agricultural Education, DECISIONS & DOLLARS, 1995

Checking accounts also have some disadvantages.

■ Some banking institutions charge a monthly fee for check processing.
■ It takes time to balance a checkbook each month.
■ Some financial institutions require that a dollar minimum be maintained.
■ If a check becomes lost or stolen, someone could gain funds at your expense.
■ There is a service charge for an overdraft.

## GUIDELINES FOR PREPARING AND COMPLETING A CHECK

Write checks legibly and in permanent ink. Including a notation stating the reason for writing the check improves efficiency in record keeping. Avoid leaving space next to the dollar sign. Checks with mistakes should be either voided or corrected and initialed (see Table 2–7).

**FIGURE 2-3**

Even young students with small supervised experience programs and little expendable cash can begin zero-balance checking accounts at some financial institutions.

## RECONCILIATION OF BANK STATEMENT

The reconciliation of bank statements verifies the current account balance. The main goals of account reconciliation are that it confirms, with bank records or the bank register, not only the transactions that have been completed and recorded by the bank, but also the account holder's transactions (Tables 2–8 and 2–9). The following steps ensure that the reconciliation is complete:

■ Mark deposits and checks that have cleared the bank.

■ Add to the balance any checks that have not cleared the bank.

■ Subtract from the balance deposits that have not cleared the bank.

■ Deduct service charges (e.g., check printing, monthly fee, and the like) from the balance.

"*T*he reconciliation process seems complicated to me," Troy remarked.

*M*s. Green found an old bank statement and reconciliation form and said, "Take a look at Tables 2–8 and 2–9. I believe after you reconcile one bank statement, you will find the process clear and logical."

**TABLE 2–7   IMPORTANT PARTS OF A CHECK***

| | No. 777 | 13-25 |
|---|---|---|
| **Gwen Green** | | 420 |
| 4215 W. Spruce Dr. | | |
| Clear Lake, OR 42156 (503) 564-2156 | | *May 20* 20 *07* |

PAY TO THE
ORDER OF _*Hoovey's Gas Station*_____   $ *24 67/100*

*Twenty-four & 67/100*_____ DOLLARS

**American Eagle Bank**
Clear Lake, OR

*Pickup repairs*_____   *Gwen Green*_____

⑆073409922⑆ 043877733 ⑈ 1988

| Check Components | Descriptions |
|---|---|
| **Gwen Green**<br>4215 W. Spruce Dr.<br>Clear Lake, OR 42156 (503) 564-2156 | Name and address of drawer may be printed on each check. |
| No. 777 | The check number is printed connectively on each check. |
| 13-25<br>420 | American Bankers Association (ABA) number. |
| *May 20* 20 *07* | The date the check was written. |
| PAY TO THE<br>ORDER OF *Hoovey's Gas Station* | Name of payee. |
| $ *24 67/100* | The check amount in Figure (numeric) format. |
| *Twenty-four & 67/100* | The check amount in word (alphabetic) format. |
| _____ DOLLARS | Line filling space between amount in word format and the word "DOLLARS." |
| **American Eagle Bank**<br>Clear Lake, OR | Bank name and location. |
| *Gwen Green* | Signature of drawer. |
| *Pickup repairs* | The ledger line is used to remind the drawer what was purchased. |
| ⑆073409922⑆ 043877733 ⑈ 1988 | Numbers printed in magnetic ink used in sorting checks. |

*Source: The National Council for Agricultural Education, DECISIONS & DOLLARS, 1995

# ■ SAVINGS ACCOUNT

Although savings accounts come in a variety of styles, they are all designed to help you save money. Because they differ in the amount of interest paid and in the accessibility of funds, one or more accounts may be appropriate for a person at any one time.

**TABLE 2-8    BANK STATEMENT***

| | | |
|---|---|---|
| 043-87773-3 | STATEMENT DATE | 3/31/07 |
| | PREVIOUS STATEMENT DATE | 2/28/07 |
| | PAGE 1 | MONTHLY STATEMENT |

*****FLAT FEE CHECKING*****

FOR DIRECT INQUIRIES OR QUESTIONS REGARDING THIS STATEMENT OR PRE-AUTHORIZED DEBITS/CREDITS—CALL CUSTOMER ASSISTANCE:

Gwen Green
4215 W. SPRUCE DR.
CLEAR LAKE, OR 42156

...............................................................................................

OR WRITE:
AMERICAN EAGLE BANK
CUSTOMER ASSISTANCE DEPARTMENT
P.O. BOX 17770
PORTLAND, OR 42377

| ACCOUNT NUMBER | PREVIOUS BALANCE | DEPOSITS & CREDITS | | CHECKS | | WITHDRAWALS & DEBITS | | CURRENT BALANCE |
|---|---|---|---|---|---|---|---|---|
| | | NO. | AMOUNT | NO. | AMOUNT | NO. | AMOUNT | |
| 043877733 | 500.00 | | | 3 | | | | 785.00 |

| LOW BALANCE (ON 03/10/07) | $441.50 | YEAR-TO-DATE INTEREST EARNED | |
|---|---|---|---|

***** DEPOSITS AND OTHER CREDITS *****

| DATE | DESCRIPTION | |
|---|---|---|
| 03/10 | DEPOSIT | |
| 03/18 | DEPOSIT | |
| 03/20 | DEPOSIT | |
| | TOTAL   3 | 360.00 |

***** CHECK TRANSACTIONS *****

| CK. NO. | DATE | TRACE NO. | AMOUNT | CK. NO. | DATE | TRACE NO. | AMOUNT |
|---|---|---|---|---|---|---|---|
| 777 | 03/03 | 00155199 | 13.00 | 779 | 03/10 | 00013777 | 5.50 |
| 778 | 03/07 | 00169422 | 40.00 | | | | |
| | | | | | TOTAL   3 | | 58.50 |

***** CASH ADVANCES, WITHDRAWALS, VISA TRANSACTIONS AND OTHER DEBITS *****

| POSTING DATE | DESCRIPTION | |
|---|---|---|
| 03/15 | PRINTING—CHECKS | |
| 03/31 | SERVICE CHARGE | |
| | TOTAL   2 | 16.50 |

*Source: The National Council for Agricultural Education, DECISIONS & DOLLARS, 1995

## TYPES

■ *Passbook* savings accounts are the standard type. The passbook is the record of account transactions. Low interest and unlimited deposits and withdrawals are passbook characteristics.

■ A *money market* account's rate of interest depends on the competitive market rates and is usually higher than that of regular savings accounts. Sometimes money market accounts have fees attached, as well as a limitation on the number of withdrawals per month.

## TABLE 2-9    BANK RECONCILIATION STATEMENT*

| BANK RECONCILIATION STATEMENT | | | |
|---|---|---|---|
| Date *3/31/07* | | | |
| Checkbook balance | *831.50* | Bank statement balance | *785.00* |
| Less service charge | *4.00* | Less outstanding checks | *20.00* |
| Less other charges | *12.50* | | |
| Deposits not credited | *50.00* | | |
| | | | |
| | | | |
| | | | |
| Adjusted balance | *765.00* | Adjusted balance | *765.00* |

*Source: The National Council for Agricultural Education, DECISIONS & DOLLARS, 1995

- *Certificates of deposit* have interest rates set by the competitive interest market. Interest rates may be fixed or variable, but they are usually higher than the current money market or regular passbook rate. Access to funds is available only after **maturity,** unless the person pays a substantial penalty. The length of deposit can vary, but it is usually one, two, or three years.

- Some financial institutions offer *youth programs* with special requirements and considerations for customers younger than 18 years of age. These programs usually have competitive interest rates with lower minimum balances.

## ADVANTAGES AND DISADVANTAGES OF SAVINGS ACCOUNTS

Savings accounts are a convenient method for accumulating funds. The deposited savings amounts earn interest, and some banking institutions offer special incentives for opening the account. The most important reason to have a savings account is that it provides reserve funds, or alternative source of funds. The disadvantages of savings accounts are that the money is not always accessible, and an absolute minimum amount of money may be required at all times. Depositing funds in a bank and withdrawing funds from a regular passbook account are simple processes when using a deposit slip (Table 2–10) and a withdrawal slip (Table 2–11).

"*I*t seems that depositing and withdrawing money is an easy process. What do you think Troy?" questioned Ms. Green.

*T*roy enthusiastically answered, "Yes, it appears easy. I am going to keep track of my spending this month and try to save enough money to start a savings account."

## TABLE 2–10   DEPOSIT SLIP*

*Deposit Ticket*

**Gwen Green**
4215 W. Spruce Dr.
Clear Lake, OR 42156
(503) 564-2156
DATE _March 20_ 20 _07_

DEPOSITS MAY NOT BE AVAILABLE FOR IMMEDIATE WITHDRAWAL

_Gwen Green_

SIGN HERE FOR CASH RECEIVED (IF REQUIRED)

**American Eagle Bank**
Clear Lake, OR 42156

|•073409922•| 043877733 ⑈• 1988

Checks and other items are received for deposit subject to the terms and conditions of this bank's collection agreement.

Be certain each item is properly endorsed. Use other side for additional listing.

| | | |
|---|---|---|
| CURRENCY | | |
| CHECKS | | |
| #0721 | 340 | 00 |
| | | |
| | | |
| | 0 | 00 |
| | | |
| | 340 | 00 |
| | 100 | 00 |
| | | |
| | 240 | 00 |

*Source: The National Council for Agricultural Education, DECISIONS & DOLLARS, 1995

## TABLE 2–11   WITHDRAWAL SLIP*

**American Eagle Bank**
Clear Lake, OR

**SAVINGS WITHDRAWAL**

043-87773-3
ACCOUNT NUMBER

**Gwen Green**
4215 W. Spruce Dr.
Clear Lake, OR 42156
(503) 564-2156

DATE _3/25/07_

RECEIVED _One-hundred and_ - - - - - - - - - - - - - - - - - - - - - - - - - - - - - - - - - - - - _no/100_ DOLLARS
(AMOUNT IN WORDS)

TO BE CHARGED TO MY ACCOUNT _$ 100.00_
AMOUNT WITHDRAWN

NEW BALANCE _$ 3,200.00_

SIGNATURE _Gwen Green_

*Source: The National Council for Agricultural Education, DECISIONS & DOLLARS, 1995

## ■ ACCOUNT MANAGEMENT

When opening a bank account, one faces a variety of account options. A good practice is to request a legal letter that describes the various account rules and regulations. This information provides the details necessary to distinguish among the options.

Today's financial world provides banking by telephone with toll-free numbers for account assistance. Money can be transferred, checks ordered, accounts opened, questions answered, and account status verified by telephone. Banking by computer is another modern option. A modem and the proper software allow individuals to bank and pay bills with a computer.

A common practice is for banks to store canceled checks rather than returning them to account holders. They make copies of checks on microfiche but provide copies to customers either free or with a charge, depending on the type of account.

Protecting an account with overdraft protection guarantees checks up to a specific dollar amount. To cover the overdraft, the bank may withdraw money from another account or send a bill to the account holders. Credit history is necessary to obtain overdraft protection.

"Insufficient funds" is the term used to describe a situation in which a bank does not cover an overdraft and returns the check to the depositor. The bank charges a substantial fee to the check writer, and the depositor (check receiver) may charge a substantial fee to the writer. The most important concern is that "bounced" checks remain on one's credit history.

Automatic withdrawal has become a popular way to pay some bills. In addition, some investments may be made through an automatic withdrawal. The account owner signs paperwork agreeing that the withdrawals may be made.

**Direct deposit** is a financial transaction in which an employer transfers earned income directly into a specified account, rather than issuing a check to the employee. The employee's bank must cooperate with the employer to accept the money transfer. Paperwork is involved only with the initial sign-up for this type of account.

An automated teller machine (ATM) makes money and other limited banking services available 24 hours a day. A personal identification number (PIN) and an ATM card are needed to operate an ATM. Interbank networks allow the use of ATM cards at banks other than your own. However, often a service charge is assessed when you use another bank's ATM machine.

## ■ SAVINGS AND INVESTMENTS

Some reasons for saving money are for retirement, a down payment (e.g., on a home or automobile), a business start-up, and education. One may invest money in many ways (Figure 2–4). **Liquidity** and risk are the two main differences among the many investment opportunities. "Liquidity" is the term used to describe the availability of the funds. **Risk** relates to the chance that the investor might lose the invested funds. The following types of personal investments are listed with the liquidity-risk factor.

**FIGURE 2-4**

Determining the right investment opportunities for each individual is made easier with accurate financial records.

| Personal Investment | Liquidity-Risk Factor |
|---|---|
| Personal loan | Not liquid, very risky |
| Savings account | Liquid, no risk |
| **Certificate of deposit** | Liquid when due, no risk |
| Money market account | Liquid, minimal risk |
| Ownership of | |
|    real estate | Not liquid, risky |
|    appreciable assets (e.g., classic cars, rare coins, etc.) | Not very liquid, risky |
| U.S. **Treasury bill** | Liquid when due, not risky |

short-term

sold at a discount (buyer earns interest through **discounting**)

maximum maturity of one year

minimum investment $10,000

| U.S. **Treasury note** | Liquid when due, not risky |
|---|---|

minimum $5,000 investment for three years or less

minimum $1,000 investment for 4 to 10 years

| U.S. **Treasury bond** | Liquid when due, not risky |
|---|---|

semiannual interest

$1,000 minimum investment

10- to 30-year maturity

| Tax-free **municipal bond** | Liquid when due, not risky |
|---|---|

issued by state, city, school district, etc.

maturity ranges from 1 to 40 years

semiannual interest

interest is tax exempt

minimum $5,000

| **Mutual fund** | Semiliquid (no due date), low risk |
|---|---|

pool of money from many individuals

invested by a professional money manager into securities

  equity mutual funds

    stock or industries

    public utilities

  fixed-income mutual funds

    corporate securities

    government agencies

  balanced mutual funds

    any combination of **stocks** and **bonds**

| **Annuity** | Some liquid, some not, low risk |
|---|---|

tax deferred

| Precious metals | Not liquid, risky |
|---|---|

gold, silver, platinum

bought and sold at **market value**

actual possession of investment

| Stock market investments | Semiliquid, risky |
|---|---|

returns to investors

  **dividend** (investors' share of annual profits)

  capital gains or losses (profits or losses from the sale of stocks)

"What is the fastest way to acquire large amounts of cash?" Troy wondered.

"I will answer that question with a question. Which would you take, a penny doubled every day for a month or $10,000 today? The answer in Table 2–12 should provide ample information to answer your question," Ms. Green said.

Troy asked, "Is that really true?"

"Yes, the key is to start saving or investing and be consistent in your efforts," retorted Ms. Green.

## ■ TIME VALUE OF MONEY

Money can grow dramatically when invested over time (creating future value), but the reverse is also true. Thus, the term **"time value of money"** is appropriate for this section (Table 2–12). **Future value** is the expected value of an account, security, investment, lump sum of money, or a series of payments, at a specific time in the future. **Present value** is the current value of one or more payments to be received or paid in the future. The present value of a future sum can appear small if the future sum is not needed for several years. The following formulas and Table 2–13 illustrate the power of **compounding** and discounting.

### FUTURE VALUE: COMPOUNDING

The formula for determining the future value (FV) of a sum is

$$FV = P(1 + I)^n$$

where $P =$ **principal,** $I =$ interest, $n =$ years. For example, if you invested $100 today at 8 percent interest for 10 years, the equation would look like this:

$$FV = 100(1 + 0.08)^{10}$$

The calculation would show that the $100 would be worth $215.89 in ten years (Table 2–13).

### PRESENT VALUE: DISCOUNTING

Likewise, when planning on saving enough money for a future purchase, you would use the present value (PV) formula:

$$PV = P\frac{1}{(1+I)^n}$$

**TABLE 2–12    PENNIES MAKE $ENSE: 1¢ DOUBLED FOR 30 DAYS\***

| Day | Beginning Value | | Ending Value |
|---|---|---|---|
| 1 | $.01 | | $.02 |
| 2 | $.02 | | $.04 |
| 3 | $.04 | | $.08 |
| 4 | $.08 | | $.16 |
| 5 | $.16 | | $.32 |
| 6 | $.32 | | $.64 |
| 7 | $.64 | | $1.28 |
| 8 | $1.28 | | $2.56 |
| 9 | $2.56 | | $5.12 |
| 10 | $5.12 | | $10.24 |
| 11 | $10.24 | | $20.48 |
| 12 | $20.48 | | $41.00 |
| 13 | $41.00 | rounded off⇒ | $82.00 |
| 14 | $82.00 | | $164.00 |
| 15 | $164.00 | | $328.00 |
| 16 | $328.00 | | $655.00 |
| 17 | $655.00 | | $1,311.00 |
| 18 | $1,311.00 | | $2,621.00 |
| 19 | $2,621.00 | | $5,243.00 |
| 20 | $5,243.00 | | $10,486.00 |
| 21 | $10,486.00 | | $20,972.00 |
| 22 | $20,972.00 | | $41,943.00 |
| 23 | $41,943.00 | | $83,866.00 |
| 24 | $83,886.00 | | $167,772.00 |
| 25 | $167,772.00 | | $335,544.00 |
| 26 | $335,544.00 | | $671,089.00 |
| 27 | $671,089.00 | | $1,342,177.00 |
| 28 | $1,342,177.00 | | $2,684,356.00 |
| 29 | $2,684,356.00 | | $5,368,711.00 |
| 30 | $5,368,711.00 | | $10,737,422.00 |

\*Source: The National Council for Agricultural Education, DECISIONS & DOLLARS, 1995

where $P$ = principal, $I$ = interest, $n$ = years. For instance, if Megan needed $2,000 for community college in five years, the PV formula could calculate the amount she would need today to reach the goal. At an 8% interest rate, the equation would look like this:

$$PV = \frac{2,000}{(1 + 0.08)^5}$$

The PV is calculated to be $1,361.17.

**TABLE 2-13    TIME VALUE OF MONEY***

Time value of money formula: $FV = P(1 + I)^n$

| | |
|---|---|
| $1,000.00 at 10% interest—1 year: | $ 1,100.00 |
| $1,000.00 at 10% interest—15 years: | $ 4,177.00 |
| $1,000.00 at 10% interest—25 years: | $10,835.00 |
| $1,000.00 at 10% interest—40 years: | $45,259.00 |

*Source: The National Council for Agricultural Education, DECISIONS & DOLLARS, 1995

The power of time coupled with interest is the basis for the financial world. Insurance companies, banks, and investment firms all utilize the basic components of compounding or discounting in their everyday transactions.

## ■ CREDIT

Most business operations are unable to grow or function year-round without borrowing money. In the short term, usually one month to a year, credit is needed to finance operating input or consumable items (e.g., fuel, fertilizer, seed, and the like). At the personal level, credit is used for everyday items (e.g., clothing, appliances, fuel, and other purchases). The credit card is a popular short-term credit instrument.

*M*egan said, "My dad believes that having credit cards is not a good idea."

"*M*egan, your dad probably is concerned when he sees people overextend their credit limit on their credit cards and then they are faced with huge monthly payments with large interest charges. It is easy to get caught up in charging things to credit cards. People need to be very careful whenever they borrow money, because they have to pay back the borrowed sums plus interest. Credit cards allow easy access to large sums of money. Paying back the borrowed money can often be difficult. However, when used wisely, credit cards can make your life easier," Ms. Green responded.

Long-term credit is required when making business or personal investments. Common examples include automobiles, equipment, and real estate (buildings and land). The usual terms for long-term credit are three to seven years for automobiles or equipment and 10 to 30 years for buildings or real estate. Credit for buildings or real estate may require a down payment (the amount of down payment varies) and may include a so-called balloon payment (i.e., a payment that is usually larger than the monthly payment and that pays off the loan).

## IMPORTANCE OF CREDIT

In the short term, credit is essential when purchasing inputs for a production cycle. This is because cash is not always available during a production cycle. Trade credit or a line of credit can be arranged through the seller. The terms can be monthly or set for the end of production cycles.

Growth and expansion require long-term credit for land and buildings. In addition, beginning a business requires an initial investment that usually means a line of credit for more than three years.

Establishing and maintaining good credit are sound agribusiness tactics. Credit bureaus provide credit history information to businesses, and merchants require this information so that they can determine how creditworthy you are. In other words, they will want to complete a financial transaction or purchase with you. The following practices build good credit history: paying bills on time; repaying loans on schedule; obtaining a regular paying job; and so on. Some financial advisors suggest that young people can establish a positive credit rating by using a credit card and paying off the charges on a regular basis. The downside to that advice is that some people don't have the self-discipline to control credit card spending. Negative credit history items include having a lien or a charge against property you own because you failed to pay a debt; adverse court judgments; an account turned over to collection agencies; and so on.

## SOURCES OF CREDIT

### Credit Cards

Credit cards are an easy source of credit. They provide an avenue of immediate payment, and they allow individuals an opportunity for purchase by telephone, mail, or Internet. In addition, there is an opportunity for instant cash (cash advance), and they can assist in building a good credit rating if managed wisely (Figure 2–5). Credit cards delay the actual cash outlay, and they provide documentation for financial transactions. Other advantages of credit cards include an opportunity for increases in credit limit and the facilitation of international travel. They are used as a fringe benefit for joining certain organizations, they are necessary when renting a car, and they are used to guarantee video rentals.

However, credit cards have several disadvantages. If a credit card is lost or stolen, someone else may gain funds at your expense. Many people overspend and are assessed extremely high interest rates (e.g., 18 percent) for using credit card funds. A monthly or yearly fee for processing may be charged, and certain businesses do not accept payment by credit card.

Common credit card companies and sources include MasterCard and VISA, with their policy of requiring a minimum payment per month. With American Express, the entire bill is due each month. Discover does not charge a fee for

**FIGURE 2-5**

Good credit allowed this student to establish a repair business while she was still in high school.

owning the card, and 1 percent of the charges becomes interest earned and is paid back to the card owner.

## Personal Loans

Operating your own agribusiness or being self-employed signifies that you are an entrepreneur. To become an agribusiness entrepreneur, one usually requires funds to get started or to expand. Chapter 10 covers loans in greater detail, but the point here is that securing personal loans is a common function among agribusiness entrepreneurs. Successful entrepreneurs become proficient at financial management skills and borrow only what is absolutely necessary according to their management plan.

The first advantage of a personal loan is that it provides large sums of money. Beyond that, personal loans assist in building a good credit rating if they are managed properly. They provide a funding source enabling individuals to advance their projects and programs. Negotiable terms and a guarantee that allows access to funds anytime can be arranged in some cases. Interest rates vary according to the institution. Personal loans can be obtained for a variety of purposes.

The reality of obtaining a personal loan is that one needs to plan for repayment according to a time line. The lending institution determines the repayment schedule and the minimum payments. Interest rates may be high depending on economic conditions. Remember that borrowed money isn't free; you pay for it. Not meeting the terms of the contract results in a poor credit rating.

**TABLE 2–14   BALANCE SHEET\***

| Assets | | Liabilities and Owner Equity | |
|---|---|---|---|
| Current assets | Value | Current liabilities | Value |
| | | | |
| Total current assets (A) | $ | Total current liabilities (D) | $ |
| Non-current assets | Value | Non-current liabilities | Value |
| Total non-current assets (B) | $ | Total non-current liabilities (E) | $ |
| | | (D + E = F) Total liabilities (F) | $ |
| | | Owner equity (C − F = G) | $ |
| (A + B = C)  Total assets (C) | $ | Total liabilities and owner equity (F + G = H) | $ |

\*Source: The National Council for Agricultural Education, DECISIONS & DOLLARS, 1995

Local banking institutions and **credit unions** are possible sources for personal loans. A **balance sheet** (the owner equity statement, Table 2–14) is often required to obtain a personal loan in addition to a personal loan application. Often, credit references and employment references are required and one's past credit history is reviewed.

Secured loans require **collateral** or a guarantee of some item of equal or greater value that will be given to the lending institution if the loan agreement is not met. Unsecured loans or signature loans are based on bank accounts, income level, and credit history, and they usually have a higher interest rate. Loan interest rates are calculated using the **annual percentage rate (APR).**

## ■ EMPLOYER EXPECTATIONS

Becoming an agribusiness employee is a logical step for many young people as they mature and attempt to become financially independent. Employers hire people to help them fulfill their agribusiness objectives. They seek people who will be productive, produce high-quality work, use good judgment, follow proper safety regulations, and take care of the physical plant, equipment, and machinery. In other words, they want to hire good citizens. Some specific attitudinal characteristics are associated with good employees and are highlighted in the next section.

### APPROPRIATE WORK HABITS

Good work habits begin with being dependable. That is, good employees arrive on time and don't miss work unless there is a real emergency. Working with others and accepting orders is the second positive work habit. Being pleasant and cooperative are habits that make going to work much more enjoyable. The third habit is to be involved with the agribusiness. Take interest and show enthusiasm for the work. Not everything in the workplace is enjoyable, but many things are, and management regards an enthusiastic employee as a person with advancement potential. Being ethical and honest comprises the fourth effective work habit. Even "borrowing" a pen that will never be returned is theft, and most agribusinesses have zero tolerance when it comes to breaking laws like stealing company property. Being proud or showing loyalty to the company is the last work habit to adopt. To be a part of the team means working out company problems or disputes in private. A positive and optimistic person who demonstrates allegiance to the agribusiness is a person who will have positive performance evaluations and a bright future. Unfortunately, unscrupulous employers do exist. When employees find themselves in situations where unethical business practices occur, they should report the situations and perhaps find other places to work.

## PERSONAL AND OCCUPATIONAL SAFETY PRACTICES

Accidents happen and the best way to prevent them is to be prepared. Employees may face many unsafe situations, such as rushing to complete a job, wearing improper clothing, working in a cluttered area, operating defective equipment, removing safety devices, following improper work procedures, lifting heavy items improperly, and so on. Develop a safety-conscious attitude.

Being a responsible employee means learning and obeying the rules and regulations associated with the agribusiness. Consulting the manual or the supervisor when a question arises is another good attribute. Watching for unsafe situations can prevent many accidents. If an accident occurs, even a small one, report it using the proper channels. Preventing future accidents by becoming involved with safety committees and improving working conditions is a very positive employee trait. Finally, follow operating procedures, and don't take shortcuts on the job.

Many organizations and agencies focus on safety. The U.S. Department of Labor offers information and workshops for agribusinesses and schools. They can provide up-to-date rules and regulations on child-labor laws, minimum wage, fair labor laws, disabilities acts, and other legal areas. Being safe and not sorry is a way of life for top-notch employees.

## Summary

Good record keeping and decision making based on sound financial management practices are lifelong skills that are important to everyone regardless of one's career. The fact that money is essential in today's world mandates an appreciation for the way money and credit work. Poor decision making at an early age can make a person a lifelong captive of high interest rates and monthly payments. Utilizing credit wisely and efficiently is a part of sound financial management practices.

### SUPERVISED EXPERIENCE

In recent years, supervised experiences have evolved to meet the needs of today's students. Figure 2–6 illustrates a typical school-based group horticulture supervised experience program in which many students operate a

**FIGURE 2-6**

Coordinating a school-based nursery operation is a complicated but rewarding task when the plants are successfully grown and ready for market.

nursery. Students in this particular program began their supervised experience by preparing a budget and an operation plan for the entire growing season.

Students detailed every step of their nursery program in their Management Information System. The following list of activities illustrates the complexity of operating a school-based nursery:

1. identifying appropriate crops for the local market (establishing demand)
2. selecting the crop(s) to be grown based on demand (the potential market), the cost of operation, and the return on investment
3. establishing an operational time line that includes student duties and responsibilities
4. implementing the activities associated with the time line
5. harvesting and selling the crop

Keeping accurate records during all phases of the school-based nursery operation was essential for determining each person's share of the profits. Students who invested more time than others were eligible to receive a greater share of the profits. Accurate records were also important when trouble, such as disease outbreaks and insect infestations, occurred. Knowing exactly when the plants were sown, transplanted, watered, and treated allowed the students to determine the most relevant solution to the problem. A portion of the profits was paid for a pizza party and movie for all the students.

## CAREER OVERVIEW

Agriscience, agribusiness, and agricultural production careers include:

High school diploma: Implement Operator; Nursery Worker

Associate degree: Florist; Biotechnology Technician

Bachelor's degree: Teacher of Agriculture; Financial Credit Specialist

Advanced degree: Financial Management Consultant; Community College Agribusiness Instructor

## DISCUSSION QUESTIONS

1. Explain the importance of personal financial management in managing and organizing personal resources.
2. Identify the advantages and disadvantages of effective personal financial management.
3. Contrast the advantages and disadvantages of checking and savings accounts.
4. List the reasons for savings and investments.
5. Compare the characteristics of various types of investments.
6. Explain the concept of interest. Describe how investments can grow in value and calculate the future value of $100 at 10 percent over five years.
7. Identify the reasons people obtain personal credit.
8. Explain the importance of the ability to obtain credit.
9. Distinguish between the types of credit. Identify possible sources of credit.
10. List the advantages and disadvantages of credit cards and personal loans. Identify the sources of personal loans.
11. Differentiate between secured and unsecured loans.

## SUGGESTED ACTIVITIES

1. Organize and categorize personal income and expenses for a period of time designated by your teacher.
2. Prepare a personal budget to accomplish given financial objectives.
3. Accurately write a check and record the transaction in a checkbook ledger (this may be a mock-up if no realistic transaction is available).
4. Accurately complete a deposit slip and a savings withdrawal slip. (This may be a mock-up if no actual transaction is available.)
5. Reconcile a bank statement and checkbook ledger. (This may be a mock-up if no actual transaction is available.)
6. Compare interest rates of various personal loans that are available in your community.
7. Obtain the information generally necessary to obtain credit cards or personal loans (use real applications if possible).

**Annual Percentage Rate (APR)**   Actual interest rate expressed on an annual basis. *Also called* effective rate.

**Annuity**   A stream of payments or dividends over time that are earned from an investment.

**Balance Sheet**   Financial statement of equity for an individual/business for a specific point in time. List of assets (current or non-current) and liabilities (current and non-current) of an individual or business. *See also* net worth statement.

$$\text{assets} = \text{liabilities} + \text{owner equity}$$

**Bond**   An interest-bearing certificate of debt to bondholder.

**Certificate of Deposit**   Special form of savings that requires a large amount of money to be invested and left in the account for a certain period of time.

**Checking Account**   Bank account against which a depositor may write checks.

**Check Register**   A separate form on which the account holder keeps a record of deposits and checks. *Also known as* checkbook ledger.

**Collateral**   Assets pledged or mortgaged to secure the repayment of a loan.

**Compounding**   Interest received from an investment. Interest is added to the principal and interest is paid again on the total sum. It is a method of calculation that can be used to determine the value.

**Credit**   Means of obtaining goods or services now by promising to pay at a later date.

**Credit Union**   Cooperative association that accepts savings deposits and makes small loans to its members.

**Direct Deposit**   Automatic deposit of paychecks into an account.

**Disbursement**   Funds paid out.

**Discounting**   Present value of a future sum accounting for the rate of compound interest. Procedure used with U.S. Treasury bills that permits buyers to purchase them at a discount and ensures a specific increase in value.

**Dividend**   Return on investment from interest and/or market appreciation.

**Future Value**   Value of an investment after it has been compounded with interest for a specified period of time (where $P$ = principal, $I$ = interest, and $n$ = years).

$$\text{FV} = P(1 + I)^n$$

**Interest**   Amount of funds paid to the lender for the use of money.

**Investment**   The outlay of money, usually for income or profit.

**Liquidity**   Consisting of or capable of ready conversion to cash. Measures the ability of a business to meet financial obligations as they come due without disrupting normal ongoing operations.

**Market Value**   The value of an asset on the open market at the present time.

**Maturity**   The point in time when investments come due and are returned to the investor.

**Money Market Account**  Savings account that pays market rate or better interest and allows access to funds without penalty.

**Municipal Bond**  Long-term, safe investment in a city, state, school district, or other governmental unit or agency.

**Mutual Fund**  Investment in which investors' money is pooled in various types of securities.

**Net Worth**  Net assets of an individual or business; owner's equity. *See also* balance sheet.

**Present Value**  The current value (PV) of an investment after it has been discounted (where $P$ = principal, $I$ = interest, and $n$ = years).

$$PV = P\frac{1}{(1+I)^n}$$

**Principal**  Amount of money borrowed or invested.

**Profit**  The difference between income and expenditures; net income.

**Risk**  The chance that the investor might lose the invested funds. Always present, it varies by degree of probability and is linked closely to feasibility and profitability. There is a risk that there may not be a profit and there is usually a risk that a plan may not be executed as intended. Risk is calculated by solvency ratios from information found in the balance sheet.

**Savings Account**  Account on which interest is paid on funds deposited by the account holder.

**Stock**  Transferable certificates representing partial ownership in a corporation.

**Tax**  Compulsory charge levied, for the common good, by a federal, state, or local unit of government against income or wealth.

**Time Value of Money**  Amount that money increases or decreases over time depending on its many alternative uses.

**Treasury Bill**  Short-term investment in the United States Treasury; sold to investors at a discount (that is, the cost is lower than the return).

**Treasury Bond**  Interest-bearing investment in the United States Treasury.

**Treasury Note**  Investment in the United States Treasury characterized by a large amount of funds, low risk, high return, and a term of one to ten years.

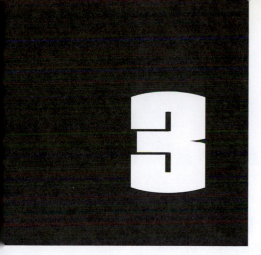

# 3 Inventory

## OBJECTIVES

*After reading this chapter, you will be able to*

- Explain the importance of an accurate inventory.
- Determine when to inventory assets.
- Identify assets to be included in the inventory.
- Given a list of assets, determine their depreciated or market value using the various methods.

## KEY TERMS

| | | |
|---|---|---|
| ACRS | Credit | Inventory |
| Acquisition | Depreciable Asset | MACRS |
| Asset | Depreciation | Non-depreciable Assets |
| Capital | Depreciation Method | Salvage Value |
| Capital Gain | Estate Tax | Straight-Line Depreciation |
| Cash Accounting | Financial Statement | Statement of Owner Equity |
| Collateral | Fiscal Year | Useful Life |
| Consumable Supplies | Income Statement | |

# BUSINESS PROFILE

Tyrone Billings was considering purchasing a bricklaying company that was on the verge of bankruptcy. The first consideration was to find out why the company was doing so poorly when so many construction jobs in the area needed skilled bricklayers. The company had a reputation for doing excellent work and was always in demand. Tyrone decided that neither the need for bricklaying nor the reputation of the company was at fault. After researching the problem for several days, Tyrone found out the company had been turned down for a loan because the lending firm decided that the company did not have enough assets to serve as collateral.

The loan was to buy supplies so that they could bid on a contract for bricking a large new grocery store that was under construction.

Tyrone asked the selling price of the company and found that the price seemed reasonable. The current owners assured him that the price included everything the business owned and gave him an inventory sheet. Tyrone was a good businessperson; so he decided to personally inventory the company to determine the sheet's accuracy. In addition to the items found on the sheet, he found that the company also owned 100,000 bricks, a cement mixer, 800 bags of cement, 86 yards of sand, a flatbed truck, a forklift, and many assorted hand tools. Because these items did not appear on the company's inventory sheet, Tyrone began to suspect that the reason for the company's financial problems was due to inaccurate bookkeeping, particularly in the area of inventory.

Tyrone bought the company and hired a person to keep an accurate accounting of all the company's assets. When filing taxes, Tyrone could accurately file the appropriate forms. Also, when he applied for a loan he had no problem in showing that the company had enough assets to cover the loan. His workers always had an adequate supply of materials and tools because he could keep supplies in stock. Without the help of an accurate inventory, Tyrone's new business would face the same financial trouble as the previous owners' business. An accurate inventory assisted him in turning the business into a profitable venture.

| Item | Quantity | Value | |
| --- | --- | --- | --- |
| | | Unit | Total |
| Cement mixer | 2 | $ 600.00 | $ 1,200.00 |
| Cement | 200 bags | $ 4.00 | $ 800.00 |
| Sand | 40 yards | $ 60.00 | $ 2,400.00 |
| Front end loader | 1 | $6,800.00 | $ 6,800.00 |
| Forklift | 1 | $7,500.00 | $ 7,500.00 |
| Scaffolding | 4 sets | $1,500.00 | $ 6,000.00 |
| Hand tools | 31 | varies | $ 4,500.00 |
| Pickup truck | 1 | $7,800.00 | $ 7,800.00 |
| | | TOTAL | $37,000.00 |

| Item | Quantity | Value | |
| --- | --- | --- | --- |
| | | Unit | Total |
| Cement mixer | 3 | $ 600.00 | $ 1,800.00 |
| Cement | 1,000 bags | $ 4.00 | $ 4,000.00 |
| Sand | 126 yards | $ 60.00 | $ 7,560.00 |
| Front end loader | 1 | $6,800.00 | $ 6,800.00 |
| Forklift | 2 | $7,500.00 | $15,000.00 |
| Scaffolding | 4 sets | $1,500.00 | $ 6,000.00 |
| Hand tools | 61 | varies | $ 8,500.00 |
| Pickup truck | 1 | $7,800.00 | $ 7,800.00 |
| Bricks | 100,000 | $ .08 | $ 8,000.00 |
| Flatbed truck | 1 | $8,800.00 | $ 8,800.00 |
| | | TOTAL | $74,260.00 |

## Introduction

The chapter explains **inventory,** which provides essential information for the four important basic financial records in the next four chapters. These chapters define and explain the balance sheet, the income statement, the statement of cash flows, and the statement of owner equity. Information contained in these **financial statements** is useful for the interpretation and understanding of the financial condition of a business enterprise.

The purpose of this chapter is to understand the importance of inventories in the financial records of a business. An understanding of correct inventory procedure is essential for accurate business management. Inventory values are an integral part of balance sheets, income statements, **credit** applications, insurance policies, estate plans, and tax management strategies. Because inventory values are such an important element of business management, a comprehensive understanding of their correct development is critical (Figure 3–1).

### ■ IMPORTANCE OF AN ACCURATE INVENTORY

Of the many uses of an inventory, six are highlighted in this section, with Troy and Megan asking questions. Many new financial terms are introduced in this chapter, and your understanding of the terms will grow as you proceed through the book. Six uses of an inventory are

1. in the owner equity statement.
2. in the income statement.
3. to obtain credit.

**FIGURE 3–1**

These two students are using a computer program to help them inventory their school-based horticulture supervised experience program.

4. for insurance purposes.

5. in estate planning.

6. for tax management.

## STATEMENT OF OWNER EQUITY

An inventory is needed to show an accurate account of the financial standing of the business at a specific point in time. Neither a **statement of owner equity** nor a balance sheet can be developed until the business inventory is complete.

> "*I*sn't all of this inventory information just counting and keeping track of what you own?" Troy questioned.
>
> *M*s. Green responded, "Troy, that is a good start on understanding inventory. I think you'll find that keeping track of what you own gets somewhat complicated when you have to determine if an item is current or non-current, depreciable or non-depreciable, but let's focus on why you inventory first."

## INCOME STATEMENT

An inventory is a necessary component when creating an **income statement.** Even if a business is using the **cash accounting** method, changes in inventory values should be included in the income statement. Identifying money brought into the business is accomplished with the help of an inventory.

## OBTAINING CREDIT

Financial institutions generally require accurate inventories as a condition before granting credit (see Figure 3–2). The inventoried **assets** may also be used as **collateral** to secure the loan.

## INSURANCE

Insurance companies require businesses to maintain accurate inventories as a basis for calculating premium costs and for paying settlements in the event of damage or loss. Inventories are needed to determine either the market or the replacement value of items in insurance claims. When purchasing insurance, consider whether the item being insured needs to be replaced with an item of the same market value or with a new item. Such decisions have a direct effect on the amount of premium.

## ESTATE PLANNING

A complete and accurate inventory is one of the first steps in estate planning and is always necessary in estate settlement situations. Again, the inventory

**FIGURE 3-2**

Keeping an accurate inventory allowed this young woman's woodworking hobby to become a viable entrepreneurial activity. Her precise records impressed the local loan officer when she applied for and received an operating loan to expand her business.

is needed to determine the value of all items that the deceased owned. Many settlement problems can be avoided and **estate taxes** can be made manageable with an accurate inventory.

> "*K*eeping an inventory is integral to the entire business operation. In fact, an accurate inventory can even affect relatives who aren't even associated with the business because of estate planning," commented Ms. Green.
>
> *M*egan added, "I think I understand. Keeping an accurate inventory will not only help you operate an efficient business, but it will also provide vital information during estate planning."

## TAX MANAGEMENT

Inventory management plays a key role in income tax management strategies, especially with cash accounting systems. Inventory control is also important in the areas of **capital gains** tax and property tax valuation.

## ■ WHEN TO INVENTORY

Inventories can and do occur throughout the year for a variety of reasons, such as determining which items to stock, preparing for a busy time of the year (e.g., purchasing enough fuel for the harvest season), deciding which input items are needed, or dropping items that are not selling well (Figure 3–3). Inventories that help owners and managers to assess the financial worth and growth of the business occur annually.

January 1 is the opening date and December 31 is the closing date of the financial year that is most often used by businesses. However, any date set by a business or organization can be used as its **fiscal year.** The important

*"If* the IRS bases its financial year on the calendar year, why would any business use a different fiscal year?" Megan asked.

*"That* is a good question. At first glance you would think that all businesses should use the calendar year, but upon closer scrutiny some other concerns emerge. Schools operate over a two-year period and use the school year for their budgeting purposes. Some operators use a cycle unique to their business; for example, some crops are planted in one year and harvested in the next year. Selecting an operating fiscal year should be based on what makes sense to the business, not because of the IRS deadlines," Ms. Green explained.

**FIGURE 3–3**

Inventory time is made easier when records are kept up-to-date during the entire year.

point is to be consistent from year to year—that is, to use the same financial period each year. Regardless of the date, the closing inventory automatically becomes the opening inventory for the new year.

New **acquisitions** must be inventoried immediately. Large items must be included in insurance policies as soon as they are purchased. It is helpful to provide the information to the insurance company prior to the actual purchase. This step allows for immediate insurance coverage upon acquisition.

## ■ WHAT TO INVENTORY

### ASSETS

All assets that will be sold during the accounting period and that are in possession at the time of the inventory should be inventoried (Figure 3–4). Agricultural assets in this category include market livestock and crops in storage.

### CONSUMABLE SUPPLIES

**Consumable supplies** that are used in the normal production cycle must be inventoried. Consumable supplies are, among other items, chemicals, fuel, replacement or repair parts, seed, feed, and fertilizer.

**FIGURE 3-4**

This student's market animal should be inventoried at the end of the fiscal year.

## CAPITAL ASSETS

An inventory of **capital** assets used by the business is important when determining business growth. Capital asset examples are animals used for breeding or replacement, machinery, equipment, tools, buildings, land, and improvements.

## ■ INVENTORY VALUES

### DETERMINING THE INVENTORY VALUE OF NON-DEPRECIABLE ASSETS

The current market value of an item is used to determine the value of **non-depreciable assets.** In other words, the inventory value is equal to the amount of cash, or the cash equivalent, that could be obtained by selling the asset. There are two categories of non-depreciable assets, current and non-current.

*T*roy stated, "I see what you mean. It does get complicated with all of the big words."

*M*s. Green said, "In essence that is what you are doing. But, to ensure that you inventory effectively and accurately, a few categories are provided to help you. I pulled out some inventory sheets that former students completed. Look at Tables 3–1 through 3–4 for an actual inventory."

Current non-depreciable assets include items that are normally turned into cash or consumed within one year. A special form (Table 3–1) is divided into current and non-current sections for proper recording of information. Current non-depreciable assets can include supplies, fertilizer, chemicals, seed, crops, feed, market livestock, and miscellaneous items. The key points are to use realistic market values and to verify that the items are consumed, sold, dispersed, or somehow used up within a year (Figure 3–5).

Non-current non-depreciable assets include items that have a **useful life** greater than one year and that have been raised or grown. The assets are not depreciable because the cost of raising or growing the items is charged as an expense. Fruit trees, Christmas trees, nut trees, raised dairy animals, draft animals, and breeding animals are all classified as non-current non-depreciable assets. These assets qualify as being grown or raised and having a useful life longer than one year.

# DETERMINING THE INVENTORY VALUES OF DEPRECIABLE ASSETS

**Depreciation** is calculated to recover the costs of "using up" assets. It is spread over the useful life of an item and is based on value. Ideally, the net book value should equal the market value. Yet many businesses depreciate their assets for tax purposes using a variety of methods that are described later in this chapter.

**TABLE 3-1 NON-DEPRECIABLE INVENTORY\***

### CURRENT NON-DEPRECIABLE INVENTORY

| Beginning Values: January 1, 2007 | | | | Ending Values: December 31, 2007 | | | |
|---|---|---|---|---|---|---|---|
| Item | Total Units | Market Value/Unit | Total Market Value | Item | Total Units | Market Value/Unit | Total Market Value |
| Market lambs | 4 | $ 60.00 | $240.00 | Market lambs | 7 | $ 60.00 | $420.00 |
| Hay (tons) | 2 | $ 90.00 | $180.00 | Hay (tons) | 1 | $100.00 | $100.00 |
| Vet supplies | — | $ 50.00 | $ 50.00 | Vet supplies | — | $ 50.00 | $ 50.00 |
| Feed (tons) | 1 | $200.00 | $200.00 | Feed (tons) | 2 | $200.00 | $400.00 |
| | | | | | | | |
| | | | | | | | |
| | | | | | | | |
| | | | | | | | |
| | | TOTAL | $670.00[1] | | | TOTAL | $970.00[2] |

### NON-CURRENT NON-DEPRECIABLE INVENTORY

| Beginning Values: January 1, 2007 | | | | Ending Values: December 31, 2007 | | | |
|---|---|---|---|---|---|---|---|
| Item | Total Units | Market Value/Unit | Total Market Value | Item | Total Units | Market Value/Unit | Total Market Value |
| Ewe lambs | 2 | $100.00 | $200.00 | Ewe lambs | 2 | $ 90.00 | $180.00 |
| | | | | Yearlings ewes | 2 | $100.00 | $200.00 |
| | | | | | | | |
| | | | | | | | |
| | | | | | | | |
| | | | | | | | |
| | | | | | | | |
| | | TOTAL | $200.00[3] | | | TOTAL | $380.00[4] |

Transfer totals [1], [2], [3], and [4] to the balance sheets.

\*Source: The National Council for Agricultural Education, DECISIONS & DOLLARS, 1995

**FIGURE 3-5**

This student is measuring the amount of soil during his end-of-the-year inventory. In this case, the greenhouse mixing soil is classified as a current non-depreciable asset.

Tables 3–2 and 3–3 include columns for both market value and adjusted cost (depreciated) value. Both methods have a useful purpose in business decision making. The market value approach provides an accurate financial picture for making decisions that affect the business operations. The depreciated value is typically used when calculating taxes. To determine the amount of depreciation to be charged, the following information must be identified for each asset:

1. original cost
2. depreciation method
3. **salvage value** (not used in most **depreciation methods**)

Types of **depreciable assets** are purchased livestock (dairy, draft, and breeding); trees (fruit, nut, Christmas, etc.) that are already in production

*M*egan queried, "If I raise an animal and keep it for breeding, then it is classified as a non-current non-depreciable asset because I charged the expense for raising it. If I purchase an animal and keep it for breeding, then it is classified as a depreciable asset. Is that correct?"

*M*s. Green enthusiastically replied, "Yes, Megan. You understand the basic concept. Of course, there are some exceptions, but most producers follow this procedure."

on purchased land; machinery; and real estate improvements. Real estate improvements consist of buildings, fences, and land improvements, such as tiling, water control structures, and ditching. Land is not depreciable, because it retains its purchase value. However, one may record market value to account for fluctuating land values.

**TABLE 3-2  DEPRECIABLE INVENTORY***

**Beginning Values**            **Date: January 1, 2007**

| Item | Date Acquired | Total Units (a) | Cost/Unit (b) | Total Acquisition Cost (a × b = c) | Depreciation (d) | Total Depreciated Value (c − d = e) | Current Market Value/Unit (f) | Total Market Value (a × f = g) |
|---|---|---|---|---|---|---|---|---|
| Purchased depreciable livestock | | | | Schedule | | | Market schedule | |
| Suffolk ram | 9/20/06 | 1 | 200 | $200.00 | $0.00 | $200.00 | $200.00 | $200.00 |
| | | | | | | | | |
| | | | | | | | | |
| | | | | | | | | |
| | | | | Subtotal (A) | | $200.00 | Subtotal | $200.00 |
| Machinery, equipment, and fixtures | | | | Schedule | | | Market schedule | |
| | | | | | | | | |
| | | | | | | | | |
| | | | | | | | | |
| | | | | | | | | |
| | | | | Subtotal (B) | | $ | Subtotal | $ |
| | | | | TOTAL (A + B = C) | | $ (5) | TOTAL | $ (5) |
| Land improvements, buildings, and fences | | | | Schedule | | | Market schedule | |
| | | | | | | | | |
| | | | | | | | | |
| | | | | | | | | |
| | | | | TOTAL (D) | | $ (6a) | TOTAL | $ (6a) |
| Land | | | | Schedule | | Total Value | Market schedule | |
| | | | | | | | | |
| | | | | | | | | |
| | | | | TOTAL VALUE (E) | | $ (6b) | TOTAL | $ (6b) |

Transfer totals (5), (6a), and (6b) to the beginning balance sheet.
*Source: The National Council for Agricultural Education, DECISIONS & DOLLARS, 1995

## TABLE 3-3    DEPRECIABLE INVENTORY*

**Ending Values**                                                    **Date: December 31, 2007**

| Item | Date Acquired | Total Units (a) | Cost/Unit (b) | Total Acquisition Cost (a × b = c) | Depreciation (d) | Total Depreciated Value (c − d = e) | Current Market Value/Unit (f) | Total Market Value (a × f = g) |
|---|---|---|---|---|---|---|---|---|
| Purchased depreciable livestock | | | | Schedule | | | Market schedule | |
| Suffolk ram | 3/20/06 | 1 | 200 | $200.00 | $50.00 | $150.00 | $200.00 | $200.00 |
| | | | | | | | | |
| | | | | | | | | |
| | | | | | | | | |
| | | | | Subtotal (A) | | $150.00 | Subtotal | $200.00 |
| Machinery, equipment, and fixtures | | | | Schedule | | | Market schedule | |
| | | | | | | | | |
| | | | | | | | | |
| | | | | | | | | |
| | | | | | | | | |
| | | | | Subtotal (B) | | $ | Subtotal | $ |
| | | | | TOTAL (A + B = C) | | $         (7) | TOTAL | $         (7) |
| Land improvements, buildings, and fences | | | | Schedule | | | Market schedule | |
| | | | | | | | | |
| | | | | | | | | |
| | | | | | | | | |
| | | | | TOTAL (D) | | $        (8a) | TOTAL | $        (8a) |
| Land | | | | Schedule | | Total Value | Market schedule | |
| | | | | | | | | |
| | | | | | | | | |
| | | | | TOTAL VALUE (E) | | $        (8b) | TOTAL | $        (8b) |

Transfer totals (7), (8a), and (8b) to the ending balance sheet.
*Source: The National Council for Agricultural Education, DECISIONS & DOLLARS, 1995

Sometimes keeping personal items separate from business items is encouraged. Table 3–4 provides an inventory form for cash, receivables, and personal assets and liabilities. Keeping personal items separate from business items allows for a more complete and accurate financial picture of a business.

**TABLE 3-4    INVENTORY OF CASH, RECEIVABLES, AND PERSONAL ASSETS AND LIABILITIES\***

| ASSETS | 1/1/07 | 12/31/07 |
|---|---|---|
| **Description** | **Beginning of Year Value** | **End of Year Value** |
| Cash on hand | $25.00 | $125.00 |
| Checking & savings accounts | | $300.00 |
| Cash value of bonds, stocks, life insurance, and co-op equities | | |
| Notes and accounts receivable | | |
| Retirement accounts (Keogh, IRA) | | |
| Other | | |
| TOTAL: Productively invested current assets | $25.00 | $425.00 |
| Apparel and jewelry | | |
| Sporting goods | | |
| Household goods | | |
| Personal share of auto/truck | | |
| Personal real estate | | |
| Other | | |
| TOTAL: Non-productively invested current assets | $ | $ |
| **LIABILITIES: PERSONAL LIABILITIES ON NON-PRODUCTIVELY INVESTED ASSETS** | | |
| Current liabilities associated with non-productively invested assets | | |
| Accounts payable | | |
| Notes | | |
| Other | | |
| TOTAL: Non-productive current liabilities | $ | $ |
| Non-current liabilities associated with non-productively invested assets | | |
| Notes of greater than 12 months | | |
| Contracts | | |
| Mortgages | | |
| Other | | |
| TOTAL: Non-productive non-current liabilities | $ | $ |

\*Source: The National Council for Agricultural Education, Decisions & Dollars, 1995

## METHODS OF DEPRECIATION

### Straight-line Depreciation

In **straight-line depreciation,** the cost of the asset minus the salvage value is depreciated equally every year of the useful life of the asset. For example, a $200 lawn mower, with a salvage value of $50 after three years, depreciates $50 per year of useful life. The purchase price of $200, less the salvage value of $50, equals $150, which is divided by three years of useful life, coming to $50 of depreciation per year.

> "*S*o, Ms. Green which method of depreciation do you recommend?" asked Troy.
>
> "*T*he important point when selecting a depreciation method is to use it consistently," commented Ms. Green. "For business decision purposes it is best to use market value. For tax purposes it is usually best to use a depreciation method for establishing value."

### Accelerated Cost Recovery System (ACRS)

An asset is depreciated more in the early years of life when using the **ACRS** depreciation method.

### Modified Accelerated Cost Recovery System (MACRS)

Similar to ACRS, an asset is depreciated more in the early years of life with the **MACRS** depreciation method. However, the years of useful life are stipulated by federal tax law. The cost of assets can be recovered over 3, 5, 7, 10, 15, 20, 27.5, or 31.5 years, depending on the type of property and the statute for that type of property.

## Summary

Inventories are important because they provide the necessary information for balance sheets, income statements, credit applications, insurance forms, estate planning, and tax management. Most businesses inventory constantly, but their financial inventory generally occurs on the fiscal year beginning with January 1 and ending with December 31. Assets are identified as being non-depreciable and depreciable. Non-depreciable assets are further divided into current and non-current categories. Several depreciation methods are described, but consistency in method is the key point.

### SUPERVISED EXPERIENCE

Inventory is extremely important for all businesses, but it takes on a special importance when a group of students attempt to run their own school-based business. This was the case with this group of girls who started an aquaculture business (Figure 3–6).

**FIGURE 3-6**

The school provided the facilities for this group's aquaculture project.

Inventory was simple when they were determining the amount of feed, counting the types of chemicals, and identifying equipment. However, they ran into major problems when it came to counting their fish. They ended up relying on the supplier for a beginning count and had to wait until harvest for a final count. The girls have a new appreciation for working together and for accurate record keeping.

## CAREER OVERVIEW

Possible agricultural careers that require financial, scientific, and production agriculture backgrounds include:

High school diploma: Irrigation Technician; Floral Trainee

Associate degree: Bank Teller; Veterinarian Assistant

Bachelor's degree: Seed Company Research Assistant; Bank Loan Advisor

Advanced degree: Seed Company Manager; University Plant Scientist

## DISCUSSION QUESTIONS

1. List the six specific uses of an inventory.
2. Select one of the uses listed in your answer to question 1 and explain why an inventory is important to it.
3. Define fiscal year.
4. When does the most prevalent fiscal year begin and end?
5. Explain why the closing inventory becomes the beginning inventory of the next year.

6. Why is it important to inventory new acquisitions immediately?

7. Distinguish among assets, consumable supplies, and capital assets.

8. Explain when livestock would be listed on the non-depreciable inventory.

9. Explain when livestock would be listed on the depreciable inventory.

10. Why are total depreciated value and total market value included on the depreciable inventory form?

## SUGGESTED ACTIVITIES

1. Inventory your personal items (use the personal inventory form, Table 3–4).

2. Inventory supplies at the school for one of the programs (e.g., greenhouse, biotechnology laboratory, aquaculture facility, livestock facility, or some other program).

3. Review inventories from former and older students who have completed comprehensive supervised experience programs.

4. Calculate the depreciated value of a five-year-old garden tractor that was purchased for $10,000 and that has a salvage value of $3,000 using straight-line depreciation method.

5. Calculate the depreciated value of the five-year-old garden tractor using another depreciation method. Describe the method and its advantages and disadvantages.

## GLOSSARY

**ACRS**   Accelerated cost recovery system, a depreciation method in which more depreciation is charged to the early years of the life of the asset.

**Acquisition**   Purchase (or receipt) of goods.

**Asset**   Anything of value owned by a business or an individual.

**Capital**   Cash or liquid savings, machinery, livestock, buildings, and other assets that have a useful life of more than one year and that meet a specified minimum value.

**Capital Gain**   Income gain that results from selling a capital asset for more than its adjusted basis (cost less depreciation).

**Cash Accounting**   Method of accounting in which income is credited to the year it was received and expenses are deducted in the year they are paid. *See also* accrual accounting.

**Collateral**   Assets pledged or mortgaged to secure repayment of a loan.

**Consumable Supplies**   Supplies and materials used in production.

**Credit**   Means of obtaining goods or services now by promising to pay at a later date.

**Depreciable Assets**   Assets that are vital to the business and that lose value over time to wear or obsolescence.

**Depreciation**   Decrease in value of business assets caused by wear and obsolescence.

**Depreciation Methods** Formula for depreciating assets (e.g., cost minus depreciation, whether by straight-line, accelerated cost recovery system, or modified accelerated cost recovery system).

**Estate Tax** Tax on the total amount of property left by a person who dies.

**Financial Statement** A statement that lists the assets and liabilities of a business at a specific time.

**Fiscal Year** Any accounting period that consists of 12 successive calendar months, 52 weeks, or 13 4-week periods. Most businesses use January 1 through December 31.

**Income Statement** A summary of revenues and expenses over a period of time.

**Inventory** Listing and valuation of all business assets.

**MACRS** Modified accelerated cost recovery system, the cost of assets recovered over 3, 5, 7, 10, 15, 20, 27.5, or 31.5 years, depending on the type of property and the statute for that type of property. For method of depreciation, *see* ACRS.

**Non-depreciable asset** An asset that does not depreciate is not kept for over one year, or is not valued at $100 or more.

**Owner Equity Statement** Financial equity statement of a business for a specific point in time. *Also called* owner equity. *See also* balance sheet.

**Salvage Value** Value of an asset at the end of its useful life (e.g., value of machinery for parts).

**Straight-line Depreciation** Depreciation method is equal to the original cost less salvage value, divided by the years of useful life.

$$SL = \frac{\text{original cost of asset } - \text{ salvage value}}{\text{useful life}}$$

**Useful Life** Number of years an asset is expected to be valuable to the business.

# Balance Sheet

*After reading this chapter, you will be able to*

- Explain the purpose of a balance sheet.
- Describe its relationship with other base financial statements.
- Describe the structure and major components of a balance sheet.
- Distinguish between assets and liabilities.
- Compare cost and market valuation of assets and liabilities.
- Give examples of each category of assets and liabilities.
- Accurately transfer financial information to a balance sheet.

## KEY TERMS

| | | |
|---|---|---|
| Asset | Equity | Owner Equity |
| Balance Sheet | Equity-to-Asset Ratio | Profitability |
| Capital | Feasibility | Repayment Capacity |
| Current Asset | Financial Efficiency | Risk |
| Current Liability | Income Statement | Solvency |
| Current Ratio | Liability | Statement of Cash Flows |
| Debt | Liquidity | Statement of Owner Equity |
| Debt-to-Asset Ratio | Non-current Asset | Working Capital |
| Debt-to-Equity Ratio | Non-current Liability | |

# BUSINESS PROFILE

*M*aria Gonzolas has saved for five years to buy her own business. She graduated from a large university with a bachelor's degree in horticulture and a minor in business administration. Her dream is to own and operate a wholesale bedding plant business. She located two different greenhouse operations for sale and is trying to determine which one is in the better financial condition. From the owners, she acquired a balance sheet that shows how well the businesses are doing. She selected the correct time to evaluate the businesses because it is the first week in January and she has the end-of-the-year balance sheets. A summary of the balance sheets for the two businesses is listed on the following page. Can you help Maria determine which business is more financially sound?

| | ASSETS | | LIABILITIES | |
|---|---|---|---|---|
| | **Business X** | **Business Y** | **Business X** | **Business Y** |
| Current | 46,321.83 | 23,243.98 | 42,876.92 | 4,879.87 |
| Non-current | 156,800.00 | 87,300.00 | 283,678.32 | 28,456.65 |
| Total | (A) $203,121.83 | (A) $110,543.98 | (B) $326,555.24 | (B) $33,336.52 |
| | Owner Equity (A − B = C) | | (C) ($123,433.41) | (C) $77,207.46 |

**SOLUTION:** Business Y is clearly in better financial condition because the assets ($110,543.98) are greater than the liabilities ($33,336.52). Not only are the liabilities ($326,555.24) of Business X greater than the assets ($203,121.83), but the current liabilities ($42,876.92) are almost as great as the current assets ($46,321.83). This could lead to financial difficulties with cash flow.

# Introduction

The purpose of this chapter is to understand a **balance sheet** (Figure 4–1) and to use it in the financial management of a business. An understanding of the components of a balance sheet is essential to effective business management. Balance sheets are an integral part of **income statements,** credit applications, insurance policies, estate plans, and tax management strategies. Because balance sheets are such an important element of business management, a comprehensive understanding of their correct development is critical.

## ■ PURPOSE OF A BALANCE SHEET

The balance sheet measures the financial condition of a business at a point in time. Specifically, the balance sheet provides an accurate picture of the financial **solvency,** or risk-bearing ability, of the business. It is important to note that the picture the balance sheet portrays does not indicate how the business achieved its financial status. The picture is simply an accumulation of the financial transactions up to the point in time when the balance sheet is prepared.

> *M*egan wondered, "Is a balance sheet really necessary?"
>
> "*A*bsolutely," responded Ms. Green. "The balance sheet has always been an essential financial statement for accountants, lenders, borrowers, operators, managers, and educators. It is one of the three base financial statements used to calculate the statement of owner equity."

The point in time that a business selects when preparing a balance sheet is important (see Figure 4–2). Although a balance sheet can be prepared at any point in time, it is recommended that it be completed at the end of the accounting period, usually December 31; the ending balance sheet becomes the beginning balance sheet for the next accounting period, which typically starts on January 1. For standardization purposes, other financial statements such as the income statement and **statement of cash flows** must be completed on the same dates as the balance sheet. Failure to follow consistent reporting dates from year to year when compiling financial statements will result in misleading information and reporting.

**FIGURE 4–1**

Balance sheet defined.

A summary of assets and liabilities (which computes owner equity) of a business at a specific point in time.

**FIGURE 4-2**

Completing financial statements at a consistent time each year is essential for accurate financial record keeping and informed decision making.

# ■ RELATIONSHIP WITH OTHER BASE FINANCIAL STATEMENTS

The balance sheet is one of the three base financial statements required to establish the **statement of owner equity** (Figure 4–3), which is the firm's check on financial record keeping accuracy at a selected point in time. The statement of owner equity verifies that the base financial statements are in agreement and that it depicts where changes in **owner equity** occur. The income statement and statement of cash flows, the other two base financial statements, are discussed in detail in the following chapters. It is important to understand that all three financial statements are related and that they are indicators of business **feasibility, risk,** and **profitability** (see Figure 4–4).

**FIGURE 4-3**

Base financial statements: The balance sheet, income statement, and statement of cash flows are the three key financial components used to calculate the statement of owner equity.

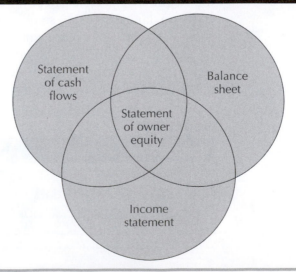

**FIGURE 4-4**

Feasibility, risk, and profitability are most closely associated with the three base financial statements.

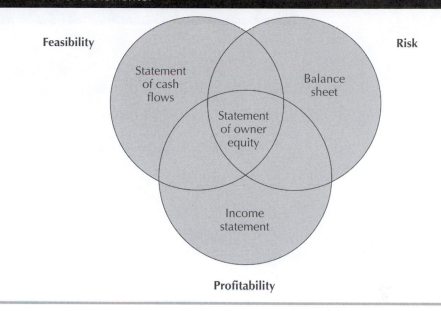

# FEASIBILITY

Feasibility is the capability of a plan to be executed successfully. Feasibility is measured by determining the **liquidity** and **repayment capacity** of a business. Liquidity measures the ability of a business to meet financial obligations as they come due without disrupting normal ongoing operations. Repayment capacity measures the firm's ability to cover term **debt** and capital lease payments. The statement of cash flows is closely related to feasibility.

# FINANCIAL RISK

Financial risk is always present, varies in its degree of probability, and is linked closely to feasibility and profitability. That is, there is a risk that there may not be a profit, and there is usually a risk that a plan may not be executed as intended. Risk is calculated by means of solvency ratios using information found in the balance sheet. Solvency describes a business that is in sound financial condition and that is able to pay all its liabilities.

# PROFITABILITY

Profitability measures the extent to which a business generates a profit or net income from the use of labor, management, and **capital.** In addition, it is the ability to retain earnings and owner equity growth. A business can survive, compete, and progress if its profitability and **financial efficiency** are at successful levels. Financial efficiency measures the degree of efficiency in using labor, management, and capital. The profitability and financial efficiency of the business (Figure 4–5) are criteria usually reflected in an income statement.

These five financial criteria—(1) liquidity, (2) repayment capacity (with feasibility), (3) solvency (with financial risk), (4) profitability, and (5) financial efficiency (with profitability)—are introduced in this chapter and explained further in Chapter 8.

**FIGURE 4-5**

Feasibility, risk, and profitability are determined by calculating the liquidity, repayment capacity, solvency, profitability, and financial efficiency of the business.

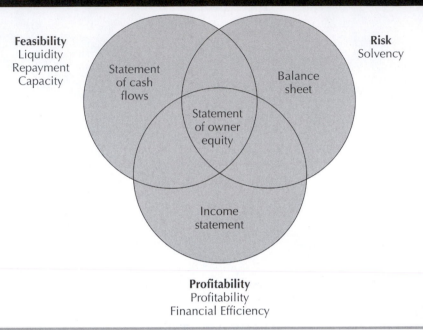

**Feasibility**
Liquidity
Repayment
Capacity

**Risk**
Solvency

Statement of cash flows

Balance sheet

Statement of owner equity

Income statement

**Profitability**
Profitability
Financial Efficiency

## ■ THE BALANCE SHEET ADDRESSES FIVE AREAS

### SOLVENCY

Do **assets** exceed liabilities? This is the solvency question, and it is essential in determining whether a business is solvent, or owns more than it owes (Figure 4–6). If a business is not solvent, it doesn't stay in operation.

### LIQUIDITY

Will the firm meet its immediate financial obligations? The balance sheet provides liquidity information that indicates whether the business can meet financial obligations as they come due without disrupting normal ongoing operations. That is, can the business pay its bills and continue its regular functions? For example, employees expect paychecks on a regular basis, and the business must have funds available for them.

**FIGURE 4-6**

The balance sheet answers five questions.

1. Do total assets exceed total liabilities (solvency)?
2. Will the firm meet immediate financial obligations (liquidity)?
3. How much loss could the business sustain (risk-bearing capacity)?
4. What collateral is available to support a loan?
5. How have business trends changed over time?

> "*A*m I able actually to project whether my business venture is worth my trouble by using a balance sheet?" asked Troy.
>
> *M*s. Green hesitated before answering, "Troy, a balance sheet is one of several financial statements needed to obtain a true and accurate picture of your business. The balance sheet can provide information on solvency, liquidity, and risk-bearing capacity. It can also determine whether you have collateral for a loan and if your business has changed over time. But you'll have to understand how the other base financial statements work together to really see if your business venture is worth your trouble."

## RISK-BEARING CAPACITY

How much loss could the business sustain? When liabilities are subtracted from assets, the net amount is an accurate figure: owner equity or net worth. (Higher net worth values are usually equated with higher risk-bearing capacity.) This risk-bearing capacity value establishes whether the business is losing or making money.

## COLLATERAL IDENTIFICATION

What collateral is available to support a loan? Inventories (see Chapter 3) are an essential ingredient when preparing a balance sheet. Therefore, when properly completed, a balance sheet identifies inventories in the form of equipment, materials, livestock, and the like—in other words, collateral that can be used to secure a loan.

## BUSINESS TRENDS

How have business trends changed over time? Evaluating financial records over a period of months and/or years is strongly encouraged. A balance sheet is a snapshot view of the business, but comparing a series of periodic balance sheets over time (e.g., at the first of each month, the end of each year, and so on) provides a long-term view that allows for informed decision making.

**FIGURE 4-7**

**Total assets always equal total liabilities plus owner equity.**

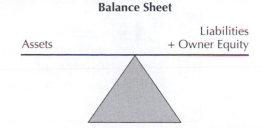

**Balance Sheet**

Assets          Liabilities + Owner Equity

> *"T*he balance sheet sort of tells a story," stated Megan.
>
> *"T*hat is correct," commented Ms. Green. "Hopefully, you understand the importance of that story. Very simply, the story is that the two sides of the balance sheet statement equal each other. Assets always equal liabilities plus owner equity. That is why it is called the balance sheet. See Figure 4–7 for a good illustration of the balance sheet concept."

## ■ STRUCTURE AND COMPONENTS OF A BALANCE SHEET

A balance sheet consists of two sides, or columns, that equal each other. One side, usually the left, contains the assets, and the other side, usually the right, contains the liabilities and owner equity. Assets and liabilities are further divided into current and non-current categories (Table 4–1).

### ASSETS

Assets are items of value owned by the business on the date the balance sheet statement is prepared. The inventory chapter provided information on identifying and valuing assets. This section explains the difference between current and non-current assets (Figure 4–8). The easier the business can turn an asset into cash, the more liquid the asset is and the higher the probability is that it will be listed as a **current asset.**

### Current Assets

Current assets are items that are used or sold and that can be converted into cash within a 12-month period without disrupting the business. Current asset examples include cash, cash value of bonds, stocks and life insurance, accounts receivable, non-depreciable items such as grain, fresh flowers, forage inventories, feeder and market livestock, and supplies such as vermiculite and fertilizer.

> *"W*hy are assets so important?" questioned Troy.
>
> *"A* good understanding of assets should help answer your question. Remember, assets include cash, supplies, and other non-depreciable items such as feed, market animals, and planting pots. An active business relies on assets for its day-to-day operations."

**TABLE 4-1   BALANCE SHEET COMPONENTS[1]\***

**Balance Sheet Statement**
**Enterprise:**                                                                                          **Date:**

| Assets | | Liabilities and Owner Equity | |
|---|---|---|---|
| Current assets | Value | Current liabilities | Value |
|  |  |  |  |
| Total current assets                    (A) | $ | Total current liabilities                    (D) | $ |
| Non-current assets | Value | Non-current liabilities | Value |
|  |  |  |  |
| Total non-current assets                (B) | $ | Total non-current liabilities                (E) | $ |
|  |  | Total liabilities          (D + E = F) | $ |
|  |  | Owner equity          (C − F = G) | $ |
| Total assets          (A + B = C) (C) | $ | Total liabilities and owner equity          (F + G = H) | $ |

[1]Assets and liabilities are categorized as either current or non-current. Total assets minus total liabilities equals owner equity.
\*Source: The National Council for Agricultural Education, DECISIONS & DOLLARS, 1995

**FIGURE 4–8**

Assets.

## ASSETS

Anything of value owned by or owed to the business on the date the balance sheet is completed.

### Current Assets

Assets used or sold and converted into cash within one year (e.g., cash, cash value of bonds, stocks and life insurance, accounts receivable, and non-depreciable items such as grain, fresh flowers, forage inventories, feeder and market livestock, and supplies such as vermiculite and fertilizer).

### Non-current Assets

Assets used in producing a product or are not sold and not converted into cash within one year (e.g., greenhouses, rototillers, buildings, machinery, irrigation systems, fences, breeding livestock, land, wells, and so on).

### Cash on Hand, Checking, and Savings

The values of these assets are straightforward accounting figures. An actual count of cash and the checking and savings account balances are included for this line on the balance sheet (Table 4–2). Also included in this category are certificates of deposit, money market accounts, and other time certificates. Again, actual values as of the date of preparation of the balance sheet are used.

### Cash Value for Bonds, Stocks, Life Insurance, and Other Investment Vehicles

These items include marketable bonds and other securities (e.g., treasury notes, government commodity certificates, government bonds, municipal securities, other bonds, and mutual funds). Both cost value with added interest or dividends and an estimated current market value are used. The difference between cost and market values is often considerable and will impact the deferred taxes column.

### Notes and Accounts Receivable

These funds are due within the year. Notes and contracts receivable are principal payments expected within 12 months. The remaining principal should

**TABLE 4-2    CURRENT ASSETS***

**Balance Sheet Statement**

Enterprise:                                    **Beginning Values:**                              **Date:**

| Assets | Value | | Liabilities and Owner Equity | |
|---|---|---|---|---|
| Current assets | Cost | Market | Current liabilities | Value |
| cash on-hand, checking and savings | | | | |
| cash value: bonds, stocks, life ins. | | | | |
| notes and accounts receivable | | | | |
| non-depreciable inventory | | | | |
| other | | | | |
| Total current assets                    (A) | $ | $ | Total current liabilities                    (D) | $ |
| Non-current assets | | Value | Non-current liabilities | Value |
| | | | | |
| | | | | |
| | | | | |
| | | | | |
| Total non-current assets                (B) | $ | $ | Total non-current liabilities                (E) | $ |

| | | | | Cost | Market |
|---|---|---|---|---|---|
| | | | Deferred taxes                    (T) | | $ |
| | Cost | Market | Total liabilities    (D + E + T = F) | $ (10) | $ (10) |
| | | | Owner equity    (C − F = G) | $ (11) | $ (11) |
| Total assets    (A + B = C) (C) | $ (9) | $ (9) | Total liabilities and owner equity    (F + G = H) | $ | $ |

*Source: The National Council for Agricultural Education, DECISIONS & DOLLARS, 1995

be listed with **non-current assets,** and the interest should be identified as accrued interest earned and listed with "other" current assets. Accounts receivable are income items related directly to the business, such as the sale of products or services that have not been paid for in cash at the time the balance sheet was completed. Feeder livestock, custom work, and crops are all examples of accounts receivable.

### NON-DEPRECIABLE INVENTORY

These inventory items were addressed at considerable length in Chapter 3. A review of the non-depreciable inventory section may be warranted.

### OTHER

Other current assets include accrued interest earned, feeder livestock and poultry on hand, crops and feed on hand, prepaid expenses (e.g., insurance, down payment, etc.), supplies, investment in a growing crop (e.g., fertilizer, seed, herbicides, beneficial insects, gas, labor, etc.), and income tax refunds.

## Non-current Assets

These assets are used in producing a product or are not sold; they are not converted into cash within one year. Converting non-current assets into cash usually has a disrupting effect on the business. Examples of non-current assets are

- non-depreciable inventory items, such as raised breeding livestock usually held for sale.
- crops held for more than a year.
- depreciable inventory items, such as rototillers, machinery, raised, and purchased breeding livestock.
- real estate and improvements, such as greenhouses, buildings, irrigation systems, fences, land, wells, and the like.
- other long term investments (Table 4–3).

Please see Chapter 3 for a review of non-current inventory assets.

### OTHER

Other non-current assets consist of some investments and the portion of notes and contracts receivable that are due beyond a year.

Outside investments are popular with many business owners. Stock obtained through a cooperative or with a credit system is listed here. Often the stock's cost and market values are the same for cooperatives and credit systems. Investments in other ventures may include business ownership and stock purchases. In either case it is imperative that both cost and market values be established.

As an example of a note or contract, the sale of real estate often results in an installment contract with the payments spread over several years. The interest and payment for the current year are included in the current assets section. The balance is listed in the non-current assets section.

## LIABILITIES

Liabilities are all the financial commitments owed by the business on the date the balance sheet statement is prepared (Figure 4–9). Loans are usually the largest **liability** category, but keep in mind that they were obtained to finance business production or to make purchases and that therefore the corresponding assets should offset the debt. Separate cost and market value columns are not required because the values would be the same—except for including deferred taxes in the market value side. This step occurs prior to adding total liabilities where the statement is divided into two columns and owner equity is determined on a cost and market value basis.

## TABLE 4-3  NON-CURRENT ASSETS*

**Balance Sheet Statement**

Enterprise:                    Beginning Values:                    Date:

| Assets | | Value | | Liabilities and Owner Equity | | |
|---|---|---|---|---|---|---|
| Current assets | | Cost | Market | Current liabilities | | Value |
| | | | | | | |
| | | | | | | |
| | | | | | | |
| | | | | | | |
| | | | | | | |
| Total current assets | (A) | $ | $ | Total current liabilities | (D) | $ |
| Non-current assets | | | Value | Non-current liabilities | | Value |
|   non-depreciable inventory | | | | | | |
|   depreciable inventory | | | | | | |
|   real estate & improvements | | | | | | |
|   other | | | | | | |
| Total non-current assets | (B) | $ | $ | Total non-current liabilities | (E) | $ |
| | | | | | Cost | Market |
| | | | | Deferred taxes (T) | | $ |
| | | Cost | Market | Total liabilities (D + E + T = F) | $ (10) | $ |
| | | | | Owner equity (C − F = G) | $ (11) | $ |
| Total assets (A + B = C) | | $ (9) | $ (9) | Total liabilities and owner equity (F + G = H) | $ | $ |

*Source: The National Council for Agricultural Education, DECISIONS & DOLLARS, 1995

## *Current Liabilities*

This debt is payable within one year. Examples of **current liabilities** are accounts payable, operating loans, accrued taxes, accrued rent, lease payments, interest on all liabilities accrued up to the balance sheet date, and principal on **non-current liabilities** due within 12 months (Table 4–4).

### ACCOUNTS AND NOTES PAYABLE

These values begin the current liabilities section. Common accounts payable entries are rent, storage, repairs, labor, feed, seed, fertilizer, chemicals, fuel, oil, and machinery hire. If the business owes money on any type of supplies, then that value becomes an account payable. Notes payable, an important component of the balance sheet, consist of only the liens that are due within a year. Only the principal portion of the note is included; the interest is included later in the accrued interest on current liabilities section.

**FIGURE 4-9**

Liabilities.

## LIABILITIES

Any financial commitment owed by the business on the date the balance sheet is completed.

### Current Liabilities

Liabilities that are payable within one year (e.g., accounts payable, operating loans, accrued taxes, accrued rent, lease payments, interest on all *liabilities* accrued up to the *balance sheet date,* principal on *non-current liabilities* due within twelve months, and so on).

### Non-current Liabilities

Liabilities due in more than one year. Formerly called intermediate (one to ten years) and long (more than ten years) term liabilities (e.g., the remaining balance of principal on equipment, machinery, breeding livestock, buildings, land, nursery stock, and so on).

"*D*o successful businesses have liabilities?" asked Megan.

"*W*hy of course they do," remarked Ms. Green. "Even if the business pays cash for all of its supplies, deferred taxes will still show up as a liability. It is very unrealistic to expect a business to operate without some type of debt."

### CURRENT PORTION OF NON-CURRENT DEBT

This section is designed to encompass all debt payments that are due within the next 12 months. For instance, if the business owes $100,000 on a 20-year loan and if $3,000 are due this year, then $3,000 is entered in the current portion of non-current debt and the remaining principal, $97,000, is entered in the non-current liabilities section for notes and chattel mortgages. Together, these two figures should always equal total debt ($3,000 + $97,000 = $100,000).

### ACCRUED INTEREST ON CURRENT LIABILITIES

This category is similar to the accrued interest section in the current assets section, but this time the interest is a debt. The term "accrue" means systematic growth. Interest grows, or accrues, on a periodic basis, usually day to day. Values placed in this section are interest charges for short-term debt, typically from accounts and notes payable.

## TABLE 4-4    CURRENT LIABILITIES*[1]

**Balance Sheet Statement**
Enterprise:                           Beginning Values:                    Date:

| Assets | | Value | | Liabilities and Owner Equity | | |
|---|---|---|---|---|---|---|
| Current assets | | Cost | Market | Current liabilities | | Value |
| | | | | accounts and notes payable | | |
| | | | | current portion of non-current debt[1] | | |
| | | | | accrued interest on current liabilities | | |
| | | | | accrued int. on non-current liabilities | | |
| | | | | other | | |
| Total current assets | (A) | $ | $ | Total current liabilities | (D) | $ |
| Non-current assets | | | Value | Non-current liabilities | | Value |
| | | | | | | |
| | | | | | | |
| | | | | | | |
| | | | | | | |
| Total non-current assets | (B) | $ | $ | Total non-current liabilities | (E) | $ |

| | | | | | Cost | Market |
|---|---|---|---|---|---|---|
| | | | | Deferred taxes    (T) | | $ |
| | | | | Total liabilities    (D + E + T = F) | $    (10) | $    (10) |
| | | Cost | Market | Owner equity    (C − F = G) | $    (11) | $    (11) |
| Total assets    (A + B = C) | | $    (9) | $    (9) | Total liabilities and owner equity    (F + G = H) | $ | $ |

[1]The current portion of non-current debt when added to the notes and chattel mortgages (total minus current portion) should equal the total debt.
*Source: The National Council for Agricultural Education, DECISIONS & DOLLARS, 1995

---

"*H*ow can interest be an asset in one case and a liability in another?" pondered Troy.

"*K*eep in mind, Troy, that interest as an asset is different from interest as a liability. For instance, if someone owes you money and you charge the person interest, then it is classified as an asset. If you owe somebody or a financial institution money and they charge you interest, then it is a liability," Ms. Green added.

### Accrued Interest on Non-current Liabilities

This balance sheet section comprises the interest to be paid in the current year on long-term debt. In the previous example, the business owed $100,000 and it paid $3,000 in principal. The interest accrued for the year was $9,000. The interest value belongs in this section.

### Other

Other current liabilities belong in this section. They may be accrued and deferred taxes, as well as loans to family members.

Personal property and real estate taxes accrue over time and can be best estimated by using last year's values plus potential changes. Income and social security taxes are other accrued taxes that may be listed if the business has unpaid taxes based on past income. This value can be obtained from the firm's tax forms, or it can be estimated based on the best available information at the time the balance sheet is prepared.

Deferred taxes on current assets means that the business is deferring, or delaying, its tax liability. Taxable income results from the sale or liquidation of all current assets. By estimating the deferred tax liability of the current assets on the appropriate worksheet, the business is counterbalancing the overstated asset values. The use of accrual accounting principals clarifies the actual business financial situation by documenting the deferred taxes in the liability sections, thus not allowing for an inflated balance sheet.

Loans to family members and friends can be listed as liabilities due to their historically low repayment rate. The term "write them off" can be applied here. If the loans are repaid, then the value is added to the current assets section. Each loan must be scrutinized with regard to its potential for repayment and then listed one way or the other: as an asset if there is a reasonable chance for repayment or as a liability if it is to be "written off."

> "**Y**ou should use caution when loaning family members or friends money because history has proven there is a very low repayment rate. This is why family and friend loans are often listed as a liability and not as an asset because the chance of their being paid back is slim," lectured Ms. Green.
>
> **M**egan's hand shot into the air, "I agree. I loaned a friend $20 last year and she moved and I never got my money back."

### Non-current Liabilities

These obligations are debt that is due in more than one year. It is important to note that intermediate- and long-term liabilities are both included as non-current liabilities (Table 4–5). Examples of non-current liabilities are the remaining principal payments on equipment, machinery, breeding livestock, buildings, land, nursery stock, and other purchases.

## TABLE 4-5  NON-CURRENT LIABILITIES*[1]

**Balance Sheet Statement**

Enterprise:                    Beginning Values:                    Date:

| Assets | | Value | | Liabilities and Owner Equity | | |
|---|---|---|---|---|---|---|
| Current assets | | Cost | Market | Current liabilities | | Value |
| | | | | | | |
| | | | | | | |
| | | | | | | |
| | | | | | | |
| | | | | | | |
| Total current assets | (A) | $ | $ | Total current liabilities | (D) | $ |
| Non-current assets | | | Value | Non-current liabilities | | Value |
| | | | | notes and chattel mortgages (total debt minus current portion)[1] | | |
| | | | | real estate mortgages/contracts (total minus current portion) | | |
| | | | | other | | |
| Total non-current assets | (B) | $ | $ | Total non-current liabilities | (E) | $ |
| | | | | | Cost | Market |
| | | | | Deferred taxes        (T) | | $ |
| | | Cost | Market | Total liabilities     (D + E + T = F) | $ (10) | $ (10) |
| | | | | Owner equity     (C − F = G) | $ (11) | $ (11) |
| Total assets | (A + B = C) | $ (9) | $ (9) | Total liabilities and owner equity  (F + G = H) | $ | $ |

[1]The current portion of non-current debt when added to the notes and chattel mortgages (total minus current portion) should equal the total debt.
*Source: The National Council for Agricultural Education, Decisions & Dollars, 1995

### NOTES AND CHATTEL MORTGAGES

These values require precise payment information for accurate data entry. The value entered here is the principal balance on the date the balance sheet is prepared. Again, using the $100,000 example, the business lists $97,000 in this section because it still owes that amount after it accounts for the $9,000 interest for the year and the $3,000 current principal payment.

Sales contracts may be listed here, especially if a sales contract is prepared. For example, you contract to deliver $10,000 worth of houseplants. The $10,000 is a liability until delivery is made, the plants are accepted, and a check for the plants is received. Until delivery, the plants serve as an asset to offset the liability of the contract.

### Real Estate Mortgages/Contracts

This balance sheet section is similar to the previous category, except that it deals only with real estate. The remaining principal is listed here, and the current payment that includes both principal and interest is listed with current liabilities.

### Other

Other non-current liabilities may, like current liabilities, include loans to family or friends and capital leases. Loans listed here are the total values that are due beyond a year and that are not listed elsewhere.

Capital leases are treated as a long-term debt. That is, the "buyout" value must be determined along with the payment schedule. The lessor can assist with determining the buyout value. The current annual payment minus interest becomes the current portion of non-current debt, the accrued interest is listed as accrued interest on non-current liabilities, and the remaining debt is listed as non-current liabilities.

### Deferred Tax Liabilities

This liability on non-current assets is calculated for the market value column on the balance sheet. Non-current asset values do not take into account the potential dollar loss from the additional taxes as a result of their sale. Therefore, the estimated income tax must be calculated and included on the liability side of the balance sheet. In the event of a business liquidation, the Internal Revenue Service tax claims can quickly absorb all liquidation cash funds. An accurate financial position statement includes deferred taxes as a part of the entire process.

## OWNER EQUITY

Owner equity is calculated by subtracting liabilities from assets (Figure 4–10), but the sources of **equity** must also be identified. For example, establishing whether owner equity was earned through business activities or accumulated from inventory inflation is extremely important.

## ■ COMPLETED BALANCE SHEETS

The ending balance sheet for one accounting year becomes the beginning balance sheet for the next accounting year. Upon the completion of a balance sheet, five financial measures can be calculated (Tables 4–6 and 4–7).

**FIGURE 4–10**

Owner equity.

**OWNER EQUITY**

value of assets
− value of liabilities
= owner equity

## TABLE 4-6  BEGINNING BALANCE SHEET*¹

**Balance Sheet Statement**
Enterprise:                          Beginning Values:                          Date:

| Assets | Value | | Liabilities and Owner Equity | |
|---|---|---|---|---|
| Current assets | Cost | Market | Current liabilities | Value |
|    cash on-hand, checking, and savings | (15) | (15) |    accounts and notes payable | |
|    cash value—bonds, stocks, life ins. | | |    current portion of non-current debt¹ | |
|    notes and accounts receivable | | |    accrued interest on current liabilities | (32a) |
|    non-depreciable inventory | (1) | (1) |    accrued int. on non-current liabilities | (16a) |
|    other | | |    other | |
| Total current assets    (A) | $ | $ | Total current liabilities    (D) | $ |
| Non-current assets | | Value | Non-current liabilities | Value |
|    non-depreciable inventory | (3) | (3) |    notes & chattel mortgages (total debt minus current portion)¹ | |
|    depreciable inventory | (5) | (5) |    real estate mortgages/contracts (total minus current portion) | |
|    real estate & improvements | (6) | (6) |    other | |
|    other | | | | |
| Total non-current assets    (B) | $ | $ | Total non-current liabilities    (E) | $ |

| | | | | Cost | Market |
|---|---|---|---|---|---|
| | | | Deferred taxes    (T) | | $ |
| | | Cost    Market | Total liabilities (D + E + T = F) | $  (10) | $  (10) |
| | | | Owner equity   (C − F = G) | $  (11) | $  (11) |
| Total assets   (A + B = C) | $  (9)   $  (9) | | Total liabilities and owner equity   (F + G = H) | $ | $ |
| Liquidity measures: #1 working capital   (A − D) | | | #2 current ratio   (A) ÷ (D) | | |
| Solvency measures: #4 debt-to-asset ratio   (F) ÷ (C) | | | #5 equity-to-asset ratio   (G) ÷ (C) | | |
| #6 debt-to-equity ratio   (F) ÷ (G) | | | | | |

Transfer values (1), (3), (5), and (6) from inventories. Transfer value (11) to ending balance sheet and to statement of owner equity. Transfer values (16a) and (32a) to income and expense summary. Transfer value (15) to statement of cash flow and financial worksheet. Transfer values (9), (10), and (11) to financial worksheet. Transfer (T) from financial worksheet.

¹The current portion of non-current debt when added to the notes & chattel mortgages (total minus current portion) should equal the total debt.

*Source: The National Council for Agricultural Education, Decisions & Dollars, 1995

**TABLE 4-7   ENDING BALANCE SHEET*[1]**

**Balance Sheet Statement**

Enterprise:                    Ending Values:                    Date:

| Assets | Value | | Liabilities and Owner Equity | |
|---|---|---|---|---|
| Current assets | Cost | Market | Current liabilities | Value |
|   cash on-hand, checking, and savings | | |   accounts and notes payable | |
|   cash value: bonds, stocks, life ins. | | |   current portion of non-current debt[1] | |
|   notes and accounts receivable | | |   accrued interest on current liabilities | (32b) |
|   non-depreciable inventory | (2) | (2) |   accrued int. on non-current liabilities | (16b) |
|   other | | |   other | |
| Total current assets          (A) | $      (17) | $      (17) | Total current liabilities          (D) | $      (18) |
| Non-current assets | | Value | Non-current liabilities | Value |
|   non-depreciable inventory | (4) | (4) |   notes & chattel mortgages (total debt minus current portion)[1] | |
|   depreciable inventory | (7) | (7) |   real estate mortgages/contracts (total minus current portion) | |
|   real estate & improvements | (8) | (8) |   other | |
|   other | | | | |
| Total non-current assets          (B) | $ | $ | Total non-current liabilities          (E) | $ |

| | | | | Cost | Market |
|---|---|---|---|---|---|
| | | | Deferred taxes          (T) | | $ |
| | | | Total liabilities     (D + E + T = F) | $      (13) | $      (13) |
| | | Cost | Market | | |
| | | | Owner equity     (C − F = G) | $      (14) | $      (14) |
| Total assets          (A + B = C) | | $      (12) | $      (12) | Total liabilities and owner equity     (F + G = H) | $      | $ |

Gain or loss in owner equity ($      (14)  − $      (11)  = I)$

| Liquidity measures:  #1 working capital          (A − D) | | #2 current ratio          (A) ÷ (D) | |
|---|---|---|---|
| Solvency measures:  #4 debt-to-asset ratio          (F) ÷ (C) | | #5 equity-to-asset ratio          (G) ÷ (C) | |
| #6 debt-to-equity ratio          (F) ÷ (G) | | | |

Transfer values (2), (4), (7), and (8) from inventories.  Transfer value (11) from beginning balance sheet. Transfer values (16b) and (32b) to income and expense summary. Transfer values (12), (13), (14), (17), and (18) to financial worksheet. Transfer (T) from financial worksheet.

[1]The current portion of non-current debt when added to the notes and chattel mortgages (total minus current portion) should equal the total debt.

*Source: The National Council for Agricultural Education, DECISIONS & DOLLARS, 1995

More financial analysis information is provided in Chapter 7, but the balance sheet measures are introduced here. These standardized measures, used to gauge financial position and strength, allow for comparisons between businesses and within a business over time (see Figure 4–11). Liquidity and solvency measures are obtained with balance sheet figures. In addition, gain or loss in owner equity can be calculated by subtracting the beginning owner equity value from the ending owner equity value (Table 4–7). If personal data is included, personal assets and liabilities are entered in the corresponding market value column. Likewise, personal net worth is included in the market column with the other owner equity values.

## LIQUIDITY MEASURES

The balance sheet's liquidity measures are limited in that they cannot evaluate whether a firm can meet all of its cash commitments. Their main purpose is to show the relationship between current assets and current liabilities. However, used in conjunction with repayment capacity measures and a cash flow budget, the balance sheet liquidity measures offer a better understanding of the financial direction of the firm.

**FIGURE 4–11**

A regular time each month should be devoted to record keeping.

> "*C*ompleting the balance sheet accurately is important because the information is used to calculate five financial measures and ratios that indicate financial position and strength. I will list the formulas as we discuss each measure," Ms. Green explained.

## Working Capital

In theory, working capital consists of the funds available to purchase business inputs and inventory after all current debts are paid and all current assets are sold. **Working capital** measures whether current assets exceed current liabilities. It is calculated by subtracting current liabilities from current assets, and it is therefore an absolute value. A business strives to maintain a positive working capital figure, but it can survive in the short term with a negative working capital if it has a steady cash flow. Business size has a tremendous impact on working capital. A $10,000 working capital figure may be adequate for a small family business, but it would be extremely low for a multimillion-dollar firm.

> "*T*o obtain financial measure number 1, the working capital measure, you subtract the total current liabilities value (D in Tables 4–6 and 4–7) from the total current assets value (A in Tables 4–6 and 4–7). This subtraction gives you the first liquidity measure," informed Ms. Green.

## Current Ratio

This ratio is a relative value calculated by dividing current assets by current liabilities. The **current ratio** indicates to what extent the current assets could pay current liabilities if they were to come due. A ratio of 1.5:1 or higher indicates that the business has enough flexibility to withstand adverse business crises such as major price changes or production catastrophes. The higher the ratio is, the more liquid the business is. It is possible to be too liquid. A firm that has a high percentage of its assets in low-yielding investments may have a high ratio, but it may be jeopardizing its objective to maximize profits.

> "*N*ow, you try it Troy. We need to find the second liquidity measure. How do you obtain financial measure number 2, the current ratio?" Ms. Green asked.
>
> "*Y*ou use the same values as the working capital measure, but now you divide the total current assets value (A in Tables 4–6 and 4–7) by the total current liabilities value (D in Tables 4–6 and 4–7)," Troy answered.

## SOLVENCY MEASURES

Solvency measures are concerned with the firm's long-term financial stability. Much like working capital as a liquidity measure, owner equity is an absolute solvency measure. Yet, as in liquidity measures, ratios provide a measure of the relationships among liabilities, assets, and owner equity that can be used for comparison between businesses and within a business over time.

### Debt-to-Asset Ratio

Dividing total liabilities by total assets is the formula for this ratio. By measuring the proportion of total assets owed to creditors, the ratio depicts the risk exposure of the firm. The firm's financial position could be at risk if the **debt-to-asset ratio** is above .50:1. The higher the ratio is, the more money creditors have in the business as compared to the owners and thus the greater the business's risk exposure is.

> "*N*ow we'll move on to solvency measures. I will do the first one, financial measure number 4," Ms. Green said. "By the way, we did not skip financial measure number 3; we simply cannot calculate it with only balance sheet information. The debt-to-asset ratio is calculated by dividing total liabilities (F in Tables 4–6 and 4–7) by total assets (C in Tables 4–6 and 4–7)."

### Equity-to-Asset Ratio

This ratio expresses the proportion of total assets financed by owner equity capital. Owner equity divided by total assets provides this financial position ratio, which is sometimes referred to as the percent ownership ratio. Therefore, the more capital that is supplied by the owner(s), the higher the **equity-to-asset ratio** is.

> *M*egan reported that, "Financial measure number 5, the equity-to-asset ratio, is computed by dividing owner equity (G in Tables 4–6 and 4–7) by total assets (C in Tables 4–6 and 4–7)."

### Debt-to-Equity Ratio

Also called the leverage ratio, the debt-to-equity ratio reflects the extent to which debt capital is being combined with equity capital. Just divide total liabilities by owner equity. Most creditors prefer that business owners have more money invested in the business than the creditors. That means that a **debt-to-equity ratio** of less than 1:1 is preferred.

*T*roy determined that, "Financial measure number 6, the debt-to-equity ratio, is found when total liabilities (F in Tables 4–6 and 4–7) is divided by owner equity (G in Tables 4–6 and 4–7)."

The accuracy of values placed on assets and liabilities is extremely important in all ratios. If a market value approach is used, then deferred taxes, cosigned notes, and other contingent liabilities must be included. However, utilizing cost values may not represent the accurate values of current assets.

## ■ COST AND MARKET VALUATION OF ASSETS AND LIABILITIES

Values must be assigned to the assets and liabilities that are identified on the balance sheet. There is no easy answer for selecting a financial value for an item (Figure 4–12). In fact, more than one method of determining financial values is necessary to provide a complete financial picture. Determining which method of valuation to use depends on how the information is to be used. Both cost values, adjusted for depreciation, and current market values should be included on the balance sheet and subsequent documents. The previous chapter described the valuation concerns when dealing with inventory items (Table 4–8). The remaining balance sheet entries are discussed in this chapter.

Inflation or deflation can create tremendous changes in the value of capital items. Utilizing both cost and market values when preparing the balance sheet provides information to determine whether changes in owner equity

**FIGURE 4–12**

Raised breeding livestock present special problems when determining their value.

## TABLE 4-8   COST AND MARKET VALUATION*[1]

| Cost Valuation | Market Valuation |
|---|---|
| **Advantages** ||
| easier to estimate accurately | easier to calculate |
| reduced fluctuations in net worth | truer representation of asset values |
| better evaluation of performance | better representation of financial position |
| **Disadvantages** ||
| original cost is difficult to determine | difficult to determine for land and buildings |
| annual record keeping is needed | difficult to determine for assets with thin markets |
| maintain two depreciation schedules (accounting vs. taxes) | adjustments must be made for deferred taxes and retained earnings vs. valuation equity |

[1]Both cost and market values are required for accurate decision making.
*Source: ES-USDA, *Looking Ahead—New Intitiatives for Financing Agriculture,* 1994

were a function of profitability (that is, they were earned) or inflation/deflation of capital assets (unearned). In addition, deferred taxes can emerge unexpectedly if both values are not utilized. Finally, cost value is required for an accrual-based income statement, and market value provides an indication of current financial status (Table 4–9).

## COST VALUATION

Cost values are usually more accurate than market values, they contribute to reduced fluctuations in net worth, and they produce a better evaluation of performance. However, original cost is difficult to determine, annual record keeping is required, and two depreciation schedules must be kept for accounting and tax purposes.

## MARKET VALUATION

Market values are easier to calculate, they furnish reliable representations of asset values, and they reflect the business's financial position better. On the other hand, it is difficult at times to determine a market value for land, buildings, and assets with slow markets. In addition, adjustments must be made for deferred taxes and retained earnings versus valuation equity.

## Summary

The balance sheet is one of three base financial statements. It systematically lists assets and liabilities at a point in time. To complete a balance sheet, identifying, listing, and valuing assets through appropriate inventory procedures is absolutely essential. Upon completion of a balance sheet, owner equity can be established and several financial ratios can be calculated for determining liquidity and solvency. Five of the sixteen ratios discussed in Chapter 8 can be computed from balance sheet figures.

**TABLE 4-9   VALUATION GUIDELINES*[1]**

| Items | Cost | Market | Other |
|---|---|---|---|
| | | **Required Disclosure** | |
| marketable securities | yes | yes | |
| inventories | no | yes | |
| PIK certificates | yes | yes | |
| accounts receivable | yes | no | |
| repaid expenses | yes | no | |
| cash investment in growing crops | yes | no | |
| purchased breeding livestock | yes | yes | |
| raised breeding livestock | recommended | yes | best method if cost is not included |
| machinery and equipment | yes | yes | |
| investments in capital leases | yes | no | |
| investment in cooperatives | no | no | net equity |
| investments in other entities | yes | no | net equity if enough ownership exists to exert control |
| cash value of life insurance | no | yes | |
| retirement accounts | yes | yes | |
| other personal assets | no | yes | |
| real estate | yes | yes | |
| buildings and improvements | yes | yes | |

[1]Consistent financial values are essential for accurate decision making.
*Source: ES-USDA, *Looking Ahead—New Intitiatives for Financing Agriculture,* 1994

## SUPERVISED EXPERIENCE

Another popular school-based supervised experience is hydroponics. Using this technique, students can grow a tremendous amount of produce in a very small area. The students in Figure 4–13 are checking the root growth on their lettuce plants. They plan to market the lettuce locally and save enough for a few meals at home. Again, accurate planning, production, and financial records are necessary to ensure that equitable profit sharing occurs upon completion of the project.

## CAREER OVERVIEW

Agricultural careers that utilize financial, scientific, and production agriculture experience include:

High school diploma: Game Farm Worker; Forestry Aide

Associate degree: Animal Health Technician; Machinery Parts Sales Associate

**FIGURE 4-13**

Hydroponics are an excellent school-based supervised experience that integrates plant science, hydrology, environmental science, and financial management.

Bachelor's degree: Financial Analyst; Landscape Architect

Advanced degree: Resource Economist; Microbiologist

## DISCUSSION QUESTIONS

1. What three base financial statements are required to prepare the statement of owner equity?

2. What is the purpose of the statement of owner equity?

3. Why is it important to complete the base financial statements on the same date?

4. Explain the relationship between the five financial criteria and feasibility, risk, and profitability.

5. Discuss the five questions that a balance sheet answers.

6. Distinguish between assets and liabilities.

7. Distinguish between current and non-current.

8. Identify three examples each of current assets, non-current assets, current liabilities, and non-current liabilities.

9. Contrast cost and market valuations of assets.

10. Illustrate the problem caused by not including deferred taxes on a balance sheet.

11. Long-term loan values are classified in three categories on the liability side of the balance sheet. Identify each category and explain which part of the loan is included in each category.

12. Identify and compare the two liquidity measures that can be generated from a balance sheet.

13. Identify and contrast the three solvency measures that can be generated from a balance sheet.

## SUGGESTED ACTIVITIES

1. Complete a balance sheet for yourself or for a school-related entity such as a club, organization, or activity.

2. Use the inventory information you collected in Chapter 3 and prepare a balance sheet.

3. Invite a representative from a financial institution to share balance sheet experiences with the class.

## GLOSSARY

**Asset**   Anything of value owned by a business or individual.

**Balance Sheet**   Financial statement of equity for an individual/business for a specific point in time. List of assets (current or non-current) and liabilities (current and non-current) of an individual or business. *See also* net worth statement.

assets − liabilities = owner equity (or net worth or net gain/loss)

**Capital**   Cash or liquid savings, machinery, livestock, buildings, and other assets that have a useful life of more than one year and that meet a specified minimum value.

**Current Asset**   An asset that is used or sold within the year.

**Current Liability**   A liability that is payable within the year.

**Current Ratio**   Extent to which current assets would cover current liabilities. It is computed by dividing current assets by current liabilities.

**Debt**   The liabilities against a person or business.

**Debt-to-Asset Ratio**   A ratio computed by dividing total liabilities by total assets. It measures the proportion of total assets owed to creditors.

**Debt-to-Equity Ratio**   A ratio that reflects the extent to which debt capital is being combined with equity capital. Divide total liabilities by owner equity. *Also called* leverage ratio.

**Equity**   Net ownership of a business. It is the difference between the assets and liabilities of an individual or business as shown on the balance sheet or financial statement.

$$\text{assets} - \text{liabilities} = \text{owner equity}$$

**Equity-to-Asset Ratio**   The proportion of total assets financed by owner equity capital. Divide owner equity by total assets. *Sometimes also called* percent ownership ratio.

**Feasibility**   The capability to execute a plan successfully. It is determined by measuring a business's liquidity and repayment capacity. The statement of cash flows is closely related to feasibility.

**Financial Efficiency**   A measurement of the degree of efficiency in using labor, management and capital.

**Income Statement**   A summary of revenues and expenses over a period of time.

**Liability**   Money, goods, and/or services that are owed.

**Liquidity**   Characteristic of an asset that makes it capable of ready conversion to cash. Measures the ability of a business to meet financial obligations as they come due without disrupting normal ongoing operations.

**Non-current Asset**   An asset that is not sold, converted into cash, or used up within the year.

**Non-current Liability**   A liability that is not due within the year. Formerly classified as intermediate (1–10 years) and non-current (more than 10 years).

**Owner Equity**   Assets minus liabilities.

**Profitability**   A measurement of the extent to which a business generates a profit or net income from the use of labor, management, and capital. It is also the ability to retain earnings and owner equity growth. A business can survive, compete, and progress if its profitability is at a successful level. The profitability and financial efficiency of the business are criteria usually found in an income statement.

**Repayment Capacity**   A measurement of the ability to cover term debt and capital lease payments.

**Risk**   The chance that the investor might lose the invested funds. Always present, it varies by degree of probability and is linked closely to feasibility and profitability. There is a risk that there may not be a profit and there is usually a risk that a plan may not be executed as intended. Risk is calculated by solvency ratios from information found in the balance sheet.

**Solvency**   Describes a business that is in sound financial condition and that is able to pay all its liabilities.

**Statement of Cash Flows**   Arranges information into operating activities, investing activities, and financing activities. The sum of the net cash flows is the change in cash during an operating period.

**Statement of Owner Equity**   The firm's check on accuracy at a point in time. It verifies that the base financial statements are in agreement and depicts where changes in owner equity occur.

**Working Capital**   The amount of funds available after sale of current assets and payment of current liabilities. Subtract current liabilities from current assets.

# Income Statement

## OBJECTIVES

*After reading this chapter, you will be able to:*

- Explain the importance of an income statement.
- Describe its relationship with other base financial statements.
- Distinguish between the cash and accrual methods of determining net income.
- Discuss the structure and major components of an income statement.
- Identify sources of revenue and expenses.
- Given a list of transactions, categorize revenues and expenses.
- Prepare and use an income statement to analyze a business.

# BUSINESS PROFILE

Colonial States Bank has been asked to issue a loan to Sleepy River Feed Company for $150,000 to install some new milling equipment. The company's old milling equipment is dilapidated and very worn. The owners think that the machinery cannot last another year and want to replace it. They explain that the mill is running about the amount of feed necessary to keep up with the demand and that the new equipment is not intended to increase the output of the mill but to maintain the current output. According to the owners, the company that manufactures the equipment estimates that the new milling equipment should last for 15 years under the conditions at Sleepy River Feed Company.

The job of the loan officers was to make sure that the feed company was making enough money to continue to operate and pay off the loan at the same time. The best way to accomplish this task was to examine the feed company's records. The bank's loan officers asked to see its income statement to help determine whether the feed mill would be able to repay the debt. The owners of Sleepy River Feed Company are good managers and they kept excellent financial records. In fact, they produced two different statements of income. One was labeled "Income Statement—Cash Accounting" and the other was labeled "Income Statement—Accrual Accounting." This made the job of the loan officers much easier.

From the cash accounting income statement the loan officers were able to determine that the company made a net profit of $226,000 last year. The accrual accounting income statement allowed the loan officers to establish the income in terms of inventories (tons of feed, grain, and so on) on hand at the end of the year, compared to the inventory at the beginning of the year. They determined that the feed company had an additional $80,000 worth of inventory on hand.

From the income statements the bank was able to determine that Sleepy River Feed Company could repay the loan over a five-year period. The loan payments each year would be $44,000 per year. When the loan payment total was subtracted from the annual income of $226,000, a profit of $182,000 was left. The bank thought the loan to the feed company was a good investment.

| | |
|---|---|
| Net profit last year | $226,000.00 |
| — Loan payment/year | $ 44,000.00 |
| Profits | $182,000.00 |

## Introduction

The purpose of this chapter is to understand the components of an **income statement** and to use it in the financial management of a business. This is the second of the four important chapters defining and explaining the base financial records: the **balance sheet,** income statement, **statement of cash flows,** and **statement of owner equity.**

## ■ IMPORTANCE OF AN INCOME STATEMENT

Determining whether a business can continue to operate is, of course, a major concern to its owners. **Profit,** or **net income,** is the most important item that is used to ascertain a firm's future. The income statement calculates the amount of a business's profit or loss (Figure 5–1), which is the principal ingredient in evaluating past, current, and future business performance and potential. The income statement information, coupled with the contents of the balance sheet, allows users to calculate several more financial ratios. Assessing a business's strengths and weaknesses is therefore enhanced by the availability of an income statement (Figure 5–2).

Historically, many businesses utilized **cash accounting** procedures when producing an income statement. In the cash method, income is credited to the year it is received and expenses are deducted in the year they are paid. Convenience and the fact that the Internal Revenue Service permits businesses to submit cash **revenue** and **expenses,** with an allowance for depreciation, did little to encourage firms to follow uniform and consistent financial practices. The cash method allowed businesses tax flexibility, which is its intent, but does not produce dependable data for financial management purposes. In **accrual accounting,** income is reported in the year earned and expenses are deducted or capitalized in the year incurred. A combination of cash and accrual accounting is necessary for firms to produce financial records that satisfy income tax concerns while providing the crucial information for determining financial analyses (Figure 5–3).

**FIGURE 5–1**

Income statement.

|  |  |
|---|---|
|  | Revenues |
| − | Expenses |
| = | Net income (profit or loss) |

**FIGURE 5–2**

The income statement answers two basic questions.

1. Did the business end the year with a profit or loss?
2. How much was the profit or loss?

**FIGURE 5-3**

Adding an accrual accounting system to the existing cash accounting system requires a combined effort from lenders and business owners.

## USES OF AN INCOME STATEMENT

There are five primary uses of an income statement (Figure 5–4). Individual revenue and expense entries are generally not included on an income statement because the report is designed for summarizing financial information. Total revenue and expense figures are calculated on the income statement. Determining profit or loss is the income statement's primary purpose. However, explaining changes in owner equity is another important use of the income statement.

**FIGURE 5-4**

Uses of an income statement.

1. Summarizing revenues and expenses
2. Determining profit or loss
3. Explaining changes in owner equity
4. Calculating financial measures
   a. profitability
   b. financial efficiency
   c. repayment capacity
5. Supporting loan application for refinancing

*T*roy queried, "Isn't making a profit the only reason to own a business?"

*M*s. Green responded, "Making a profit is extremely important in a business, but there are times when a business can survive, usually for a short period of time, without making a profit. Understanding if the business has a feasible chance to recover and repay debts is made possible if the owner has prepared a sound set of financial statements. Mr. Browning, the owner of Browning and Associates, a local produce clearinghouse, will be by later to visit with me about hiring students for summer internships. Let's ask him about his financial records."

Three financial criteria are addressed with the income statement: profitability, financial efficiency, and repayment capacity. Finally, the income statement provides support for loan applications.

## ■ RELATIONSHIP WITH OTHER BASE FINANCIAL STATEMENTS

Financial reporting includes information contained in financial statements as well as information considered necessary or useful in the interpretation and understanding of the financial condition of a business enterprise (Figure 5–5). The income statement, as one of the four base financial statements, is integral for evaluating the firm's financial performance and making management decisions. It should be completed for the same time period as that of the

**FIGURE 5–5**

The integration of the four base financial statements provides a complete picture of a business's financial position.

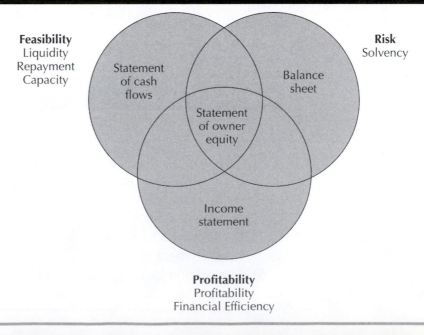

**Feasibility**
Liquidity
Repayment
Capacity

Statement of cash flows

Statement of owner equity

Balance sheet

**Risk**
Solvency

Income statement

**Profitability**
Profitability
Financial Efficiency

balance sheet and the other financial statements. This time period often corresponds with the income tax reporting year.

## ■ CASH AND ACCRUAL ACCOUNTING

The basic difference between cash and accrual accounting lies in the approach each system uses to identify **income** and expenses. Cash accounting recognizes income and expenses when the income is actually received and when the expenses are actually paid. Accrual accounting recognizes any transaction that increases owner equity as income and any transaction that decreases owner equity as expenses. In addition, accrual accounting records income and expenses when they are incurred rather than when they are received or paid (Tables 5–1 and 5–2).

### TABLE 5–1   INCOME STATEMENT: CASH*

**Name** _____        **12-month period ending** _____

| Revenue | | Cash |
|---|---|---|
| Cash revenue | | |
| | $ | |
| | Gross cash revenues | $ |
| | | |
| | Gross revenues | = $ |
| **Expenses** | | |
| Cash expenses | | |
| | $ | |
| | Total cash expenses | − $ |
| | Net cash income | = $ |
| | Depreciation | − $ |
| | | |
| | Net income from operations | = $ |
| | Gain/loss on sale of capital assets | ± $ |
| | Net income | = $ |

*Source: The National Council for Agricultural Education, Decisions & Dollars, 1995

## TABLE 5-2    INCOME STATEMENT: ACCRUAL*

Name _____          12-month period ending _____

| Revenue | | | | Accrual |
|---|---|---|---|---|
| Cash revenue | | | | |
| | | $ | | |
| | | | Gross cash revenues | $ |
| Inventory adjustments | **Inventories** Beginning | Ending | **Difference** (End. – Beg.) | |
| | $ | $ | $ | |
| | | | Total inventory adjustment | ± $ |
| | | | Gross revenues | = $ |
| **Expenses** | | | | |
| Cash expenses | | | | |
| | | $ | | |
| | | | Total cash expenses | – $ |
| | | | Net cash income | = $ |
| Depreciation | | | | – $ |
| **Noncash expense adjustment** | | | | |
| Assets | **Accounts** Beginning | Ending | **Difference** (End. – Beg.) | |
| Unused supplies Other | $ | $ | $ | |
| Liabilities | | | | |
| Accounts payable Accrued interest Accrued taxes Other | $ | $ | $ | |
| | | | Total noncash expense adjustments | ± $ |
| | | | Net income from operations | = $ |
| | | | Gain/loss on sale of capital assets | ± $ |
| | | | Net income | = $ |

*Source: The National Council for Agricultural Education, DECISIONS & DOLLARS, 1995

"*M*r. Browning, which method of accounting do you use with your produce clearinghouse, cash or accrual?" Troy asked.

"*T*he question is not which method do I use, but when do I use each method. Very simply, I use the accrual method to make sound financial decisions about my business. I use the cash method when I am preparing my taxes," Mr. Browning replied. "I believe that Ms. Green is going to explain in her next class the pitfalls of using only cash accounting," Mr. Browning added.

As in all of the base financial statements, accuracy is critical when preparing an income statement. The example in Figure 5–6 illustrates the problem of not being accurate. It is not unusual for people to overestimate revenue by 15 percent and underestimate expenses by 17 percent when using cash accounting. A 303 percent error occurs in this example as a result of inaccuracy. The important lesson is to complete the forms as accurately as possible or don't do it at all, especially when creating the data that generates financial ratios. Doing it right means using cash and accrual accounting methods. Cash methods are used for income tax preparation, and accrual methods are used for analyzing the financial performance of the business.

Other pitfalls that may occur if cash accounting is the only method used include the use of several non-recommended techniques that give the appearance of profitability. The following techniques are usually employed when a business is undergoing financial difficulty and its owners are using stopgap measures to keep the business afloat:

- selling inventory
- selling capital assets

**FIGURE 5–6**

Inaccurate figures in an income statement can be devastating.

| | Estimated | | Transaction | Actual |
|---|---|---|---|---|
| | **Estimated versus Actual Values** | | | |
| | Estimated | | Transaction | Actual |
| Revenue | 100,000.00 | overestimated by 15% | subtract 15,000.00 | |
| Expenses | 90,000.00 | underestimated by 17% | add 14,450.00 | |
| Profit | 10,000.00 | | | |
| Loss | | | | |
| Error | 29,400.00 10,000.00 | difference (projected profit minus actual loss) projected profit | | |

- living off depreciation
- accruing expenses
- accumulating **accounts payable**
- refinancing operating losses

These steps often give the appearance of a prosperous business, but if items like deferred taxes and other noncash entries are not taken into account, it becomes clear that the owners are only deceiving themselves. However, it is important to note that business growth and expansion can also be concealed with cash basis reporting. Regardless of the method used, revenue and expenses are entered on the statements.

## ■ REVENUE

Two main categories of revenue encompass the accrual income statement, cash sales and inventory adjustments. Cash sales are usually categorized by enterprises such as bedding plants, crops, or marketable fish. Custom work or services performed are entered in the revenue section. Noncash entries include inventory adjustments, capital gain or loss, and accounts receivable (Figures 5–7 and 5–8).

> "*H*ow do you remember all of the figures when you are preparing your income statement?" Megan questioned.
>
> "*I* don't remember any figures. I record all of my financial entries as they occur in my financial ledger, then I summarize and calculate the totals on special forms in my management information system," Mr. Browning replied.

### INCOME AND EXPENSE SUMMARY SHEETS

Prior to completing the income statement, the balance sheets and the income and expense summary sheets must be completed. Making adjustments to income and expenses, as a result of employing accrual accounting, is the primary purpose of the income and expense summary sheets. The first income and expense summary sheet (Table 5–3) provides totals from the record of income and expenses. Each income summary sheet (see Table 5–9 on page 117) and expense summary sheet (see Table 5–10 on page 119) entry is coded (see Table 5–6 on page 111), and, when needed, the totals are calculated on the summary sheets.

The income and expense summary schedules (Table 5–4) utilize data from the balance sheets and from other summaries, and they produce figures essential for the income statement and other financial statements. Specifically,

**FIGURE 5-7**

Revenue.

Cash examples of revenue:
1. Cash sales
   a. plants/crops
   b. manufacture goods
   c. livestock
2. Government payments
3. Custom work receipts

Noncash examples of revenue (used in accrual accounting):
1. Changes in accounts receivable
2. Changes in inventories
   a. crops/plants
   b. livestock
   c. supplies

**FIGURE 5-8**

Accrual accounting includes increases in crop inventories as noncash sources of revenue.

interest expenses, gifts and inheritances, and additions to paid-in capital are calculated in these schedules.

The final income and expense summary calculates the gain or loss on the sale of capital assets (Table 5–5). This schedule adjusts capital sales so that they take into account differences from beginning and ending inventory values and differences in value as a result of depreciation.

## TABLE 5-3   INCOME AND EXPENSE SUMMARY*

**Enterprise** _____

Add all income and expense items with the same code and record on this page. Transfer the totals to the income statements, statement of cash flows worksheet, and the statement of cash flows.

Note: For analysis purposes, the previous third and fourth level categories (e.g., 2.1.1, 2.1.4.1, and so on) have been combined to the larger categories listed below. The third and fourth level categories needed for financial analysis are calculated on the financial schedules that follow.

1 _cash withdrawals for family living:_ ...................................................................................... (35) $_____ E

2 _cash paid for operating activities:_

    2.1  cash paid for operating activities: feeder animal, feed purchases, operating and interest expenses, and hedging account deposits ............................................ $_____ E

    2.2  cash expenses paid in nonbusiness operations ............................................... $_____ E

    2.3  income and social security taxes paid in cash .............................................. $_____ E

    2.4  extraordinary items paid in cash ................................................................ $_____ E

    2.5  cash withdrawal for family living ............................................................... $_____ E

    2.6  cash withdrawals for investment into personal assets ..................................... $_____ E

3 _cash received from operating activities:_

    3.1  cash received from business operations (all cash sales from income statement)....... $_____ R

    3.2  cash received from nonbusiness income and operations ................................... $_____ R

    3.3  extraordinary items received in cash .......................................................... $_____ R

4 _cash paid for investing activities:_

    4.1  breeding & dairy livestock ......................................................................... $_____ E

    4.2  machinery & equipment ............................................................................ $_____ E

    4.3  business real estate; other business assets .................................................. $_____ E

    4.4  capital leased assets ............................................................................... $_____ E

    4.5  bonds & securities; investments in other entities; other nonbusiness assets .......... $_____ E

5 _cash received from investing activities:_

    5.1  raised breeding & dairy livestock (not capitalized & not depreciated) ................... $_____ R

    5.2  purchased & raised breeding & dairy livestock (capitalized & depreciated) ............ $_____ R

    5.3  machinery & equipment ............................................................................ $_____ R

    5.4  business real estate; other business assets .................................................. $_____ R

    5.5  bonds & securities; investments in other entities; other nonbusiness assets .......... $_____ R

6 _distribution of dividends, capital, or gifts:_

    6.1  cash portion............................................................................................ $_____ R/E

    6.2  property portion ...................................................................................... $_____ R/E

7 _cash received from equity contributions:_ ................................................................ $_____ R

    7.1  gifts & inheritances.................................................................................. $_____ R

    7.2  additions to paid-in capital including investments of personal assets into the business................................................................................................. $_____ R

    7.3  investments of personal assets into the business........................................... $_____ R

8 _principal paid on term debt & capital leases:_

    8.1  term debt principal payments: scheduled payments........................................ $_____ E

    8.2  term debt principal payments: unscheduled payments..................................... $_____ E

    8.3  principal portion of payments on capital leases.............................................. $_____ E

9 _operating loans received (including interest paid by loan renewal):_ ......................... (53) $_____ R

10 _proceeds from term debt (term debt financing; loans received):_........................... (45) $_____ R

11 _operating debt principal payments:_ ...................................................................... (54) $_____ E

E: indicates expenses; R: indicates revenue; R/E indicates either expense or revenue.

Transfer values (35), (53), (45), and (54) to statement of cash flows.

*Source: The National Council for Agricultural Education, DECISIONS & DOLLARS, 1995

**TABLE 5-4  INCOME AND EXPENSE SUMMARY—SCHEDULES***

**Enterprise** _____

**Interest Expense Schedule**

| | |
|---|---|
| interest paid in cash or by renewal on operating loans & accounts payable:<br>Total of code # 2.1.4.1 = (A) | (33) |
| interest paid in cash or by renewal on term debt & capital leases:<br>Total of code # 2.1.4.2 = (B) | (22) |
| Total interest paid in cash or by renewal (A + B = C)(C) | $ |

| Transfer values for [(32a), (16a), (32b), and (16b)] from balance sheets. | | beginning of period | end of period | adjustments (+ or −) |
|---|---|---|---|---|
| accrued interest on current liabilities: | $[^{(32b)} - {}^{(32a)} = {}^{(32)}]$ | (32a) | (32b) | ± (32) |
| accrued interest on non-current liabilities: | $[^{(16b)} - {}^{(16a)} = {}^{(16)}]$ | (16a) | (16b) | ± (16) |
| | $[^{(32a)} + {}^{(16a)} = {}^{(34a)}]$ and $[^{(32b)} + {}^{(16b)} = {}^{(34b)}]$ | (34a) | (34b) | |
| | $[^{(34a)} + {}^{(34b)} = {}^{(34)}]$ | | (34) | |
| Total change in accrued interest payable $[^{(32)} + {}^{(16)} = D]$ | | | | $ |

Transfer values [(34a), (34b), and (34)] to income statement: accrual.
Transfer values [(16), (22), and (34b)] to financial worksheet.


**Gifts & Inheritance Schedule**

| | |
|---|---|
| cash portion of gifts & inheritances:<br>Total of code # 7.1.1 = (A) | (41a) |
| property portion of gifts & inheritances:<br>Total of code # 7.1.2 = (B) | (41b) |
| Total gifts & inheritances (A + B = C)(C) | (41) |

**Additions to Paid-In Capital Schedule**

| | |
|---|---|
| cash portion of additions to paid-in capital including investments of personal assets into the business:<br>Total of code # 7.2.1 = (A) | (42a) |
| property portion of additions to paid-in capital including investments of personal assets into the business:<br>Total of code # 7.2.2 = (B) | (42b) |
| Total additions to paid-in capital (A + B = C)(C) | (42) |

Transfer values [(41a) and (42a)] to statement of cash flows worksheet.
Transfer values [(41) and (42)] to statement of owner equity.
*Source: The National Council for Agricultural Education, DECISIONS & DOLLARS, 1995

# REVENUE ENTRIES

The revenue categories found on the record of income and expense code sheet (Table 5–6) should be used on the income statement. Customizing the subcategories for a specific business allows the owner(s) to be more efficient when it is time to analyze the firm. Also called a "chart of accounts," the code

**TABLE 5-5    INCOME AND EXPENSE SUMMARY: GAIN/LOSS ON SALE OF CAPITAL ASSETS\*\***

Enterprise _____

| Capital asset account adjustment for sales and transfers of raised breeding and dairy livestock (not fully capitalized and not depreciated) | | | |
|---|---|---|---|
| | beginning inventory | ending inventory | adjustment (+ or −) |
| value of raised breeding and dairy livestock — use column e or g | −$ | +$ | ±$ |
| cash sales of raised breeding and dairy livestock . . . . . . . . . . . . . . . . . . . . . . . . . . . . . . . . . . . | | + | (48) |
| net capital asset account adjustment (A) | | | ±$ |

| Capital gain (loss) on breeding and dairy livestock (fully capitalized and depreciated) | | | |
|---|---|---|---|
| Sales of: | net sales amount | adjusted cost/ basis\* | gain (loss) (+ or −) |
| purchased and raised breeding and dairy livestock . . . . . . . . . . . . . | +$ (49) | −$ | ±$ |
| death loss or other casualty loss . . . . . . . . . . . . . . . . . . . . . . . . | +$ | −$ | ±$ |
| total gain or loss (B) | | | ±$ |

| Capital gain (loss) on machinery, real estate, and other assets | | | |
|---|---|---|---|
| Sales of: | net sales amount | adjusted cost/ basis\* | gain (loss) (+ or −) |
| machinery and equipment . . . . . . . . . . . . . . . . . . . . . . . . . . . . . . | +$ (50) | −$ | ±$ |
| business real estate and other business assets . . . . . . . . . . . . . . . | +$ (51) | −$ | ±$ |
| total gain or loss (C) | | | ±$ |
| Gain/loss on sale of capital assets (A + B + C = D) | | | ±$ (52) |

\*Usually the adjusted cost/basis is the most current value from column e or g on the inventory sheets.

Transfer values (48), (49), (50), and (51) from statement of cash flows worksheet.

Transfer value (52) to income statements.

\*\*Source: The National Council for Agricultural Education, Decisions & Dollars, 1995

sheet varies from one business to another in its numbering and specialized categories. Table 5–6 is a standard chart that can be adapted to all businesses. It is essential that every income and expense entry (see Tables 5–9 and 5–10) is coded so that the financial analysis is streamlined and the data can be computer sorted.

## ACCOUNTS RECEIVABLE

Accounts receivable is a rather new area in the income statement revenue section for some small businesses. As businesses move toward an increasingly standard way of conducting financial analyses, including uniform records will be the norm. This entry reflects sales or services rendered for which the business has not actually received payment (Tables 5–7 and 5–8).

## TABLE 5-6    RECORD OF INCOME AND EXPENSE CODE SHEET*

**The following items are coded and the code number should be included in the code column on the following "Record on Income and Expenses" pages. You may include additional codes if you desire.**

1 *cash withdrawals for family living:*
  1.1 cash withdrawal for family living
  1.2 cash withdrawals for investment into personal assets
2 *cash paid for operating activities:*
  2.1 cash paid for operating activities
    2.1.1 feeder animal
    2.1.2 feed purchases
    2.1.3 operating expenses—note you may want to further categorize this item such as: 2.1.3.1—fuel; 2.1.3.2—oil/lubricants; 2.1.3.3—fertilizer; and so on.
    2.1.4 interest expenses—paid in cash or by loan renewal
      2.1.4.1 operating loans and accounts payable[33]
      2.1.4.2 term debt and capital leases[22]
    2.1.5 hedging account deposits
  2.2 cash expenses paid in nonbusiness operations
  2.3 income and social security taxes paid in cash
  2.4 extraordinary items paid in cash
3 *cash received from operating activities:*
  3.1 cash received from business operations (all cash sales from income statement)
    3.1.1 feeder livestock/poultry sales
    3.1.2 crops/feed
    3.1.3 custom work
    3.1.4 government payments
  3.2 cash received from nonbusiness income and operations
    3.2.1 wages
    3.2.2 interest
  3.3 extraordinary items received in cash
4 *cash paid for investing activities:*
  4.1 breeding and dairy livestock
  4.2 machinery and equipment
  4.3 business real estate; other business assets
  4.4 capital leased assets
  4.5 bonds and securities; investments in other entities; other nonbusiness assets

5 *cash received from investing activities:*
  5.1 raised breeding and dairy livestock (not capitalized and not depreciated)
  5.2 purchased and raised breeding and dairy livestock (capitalized and depreciated)
  5.3 machinery and equipment
  5.4 business real estate; other business assets
  5.5 bonds and securities; investments in other entities; other nonbusiness assets
6 *distribution of dividends, capital, or gifts:*
  6.1 cash portion
  6.2 property portion
7 *cash received from equity contributions:*
  7.1 gifts & inheritances[41]
    7.1.1 cash portion[41a]
    7.1.2 property portion[41b]
  7.2 additions to paid-in capital including investments of personal assets into the business[42]
    7.2.1 cash portion[42a]
    7.2.2 property portion[42b]
  7.3 investments of personal assets into the business
8 *principal paid on term debt and capital leases:*
  8.1 term debt principal payments: scheduled payments
  8.2 term debt principal payments: unscheduled payments
  8.3 principal portion of payments on capital leases
9 *operating loans received (including interest paid by loan renewal):*
10 *proceeds from term debt (term debt financing; loans received):*
11 *operating debt principal payments:*

Note: It is important to record noncash items (such as: receiving pasture in exchange for labor, no cash was involved, but a value should be established on both pasture and labor). Include them as you would cash items, place a value on the items and identify them with a code. You may want to separate cash and noncash items during your analysis.
Note: Values [22], [33], [41], and [42] are calculated on the income and expenses summary schedules.
*Source: The National Council for Agricultural Education, DECISIONS & DOLLARS, 1995

## ■ EXPENSES

Expenses are divided into operating and financial expenses or into cash and noncash expenses. This step allows for more useful information for financial analyses. Operating expense efficiency can be compared without the distraction of different debt loads, interest payments, and capital adjustments (Figure 5–9).

Cash operating expenses include values paid for chemicals, fuel, seeds, and other items related to running the business. Depreciation and adjustments in accounts payable, accrued expenses (such as **accrued interest** and **accrued taxes**), prepaid expenses, cash invested in growing crops, and **unused supplies** are noncash entries (Figure 5–10).

## TABLE 5–7  INCOME STATEMENT CATEGORIES: CASH*

Enterprise _____          12-month period ending _____

| Revenue | | Cash |
|---|---|---|
| Cash revenue from income and expense summary | | |
| 3 *cash received from operating activities:*  $<br>  3.1  cash received from business operations<br>      3.1.1  feeder livestock/poultry sales<br>      3.1.2  crops/feed<br>      3.1.3  custom work<br>      3.1.4  government payments | | |
| | Gross cash revenues (A) | $     (25) |
| **Expenses** | | |
| Cash expenses from income and expense summary | | |
| 2 *cash paid for operating activities:*  $<br>  2.1  cash paid for operating activities<br>      2.1.1  feeder animal<br>      2.1.2  feed purchases<br>      2.1.3  operating expenses<br>          2.1.3.1  fuel<br>          2.1.3.2  oil/lubricants<br>          2.1.3.3  fertilizer<br>      2.1.4  interest expenses<br>      2.1.5  hedging account deposits | | |
| | Total cash expenses (B) | $     (26) |
| | Net cash income (A − B = C)(C) | $ |
| | Depreciation expense from inventory (D) | $     (27) |
| | Net income from operations (C − D = E) | $     (28) |
| | Gain/loss on sale of capital assets (F) | $     (52) |
| | Net business income (E − F = G) | $     (29) |
| | Nonbusiness income (H) | $     (30) |
| | Net income (G + H = I) | $ |

Transfer value [52] from income and expense summary and value [30] from SCF worksheet.
Transfer totals [25], [26], [27], [28], and [30] to financial worksheet and [29] to SOE.
*Source: The National Council for Agricultural Education, Decisions & Dollars, 1995

## TABLE 5-8   INCOME STATEMENT CATEGORIES: ACCRUAL*

**Enterprise** _____          **12-month period ending** _____

| Revenue | Accrual |
|---|---|
| Cash revenue from income and expense summary | |

| 2 *cash received from operating activities:* $ | |
|---|---|
|    2.1  cash received from business operations | |
|       2.1.1  feeder livestock/poultry sales | |
|       2.1.2  crops/feed | |
|       2.1.3  custom work | |
|       2.1.4  government payments | |
|    2.2  cash received from nonbusiness income and operations | |
|    2.3  extraordinary items received in cash | |
| *accounts receivable* | |

| | | Gross cash revenues (A) | $ |
|---|---|---|---|

| Inventory adjustments | Inventories Beginning | Ending | Difference (End. − Beg.) |
|---|---|---|---|
| breeding/dairy livestock<br>machinery and equipment<br>real estate<br>other | $ | $ | $ |

| | Total inventory adjustment (B) | $ | |
|---|---|---|---|
| | Gross revenues (A ± B = C)(C) | $ | (25) |

| **Expenses** | |
|---|---|
| Cash expenses from income and expense summary | |

| 2 *cash paid for operating activities:* $ | |
|---|---|
|    2.1  cash operating activities | |
|    2.2  cash expenses paid in nonbusiness operations | |
|    2.3  income and social security taxes paid in cash | |
|    2.4  extraordinary items paid in cash | |
|    2.5  cash withdrawals for family living | |
|    2.6  cash withdrawals for investment into personal assets | |
| *income tax expense from statement of cash flows worksheet* | |

| | Total cash expenses (D) | $ | (26) |
|---|---|---|---|
| | Depreciation expense from inventory (E) | $ | (27) |

| **Noncash expense adjustments** | | | |
|---|---|---|---|
| Assets from balance sheets | Accounts Beginning | Ending | Difference (End. − Beg.) |
| Unused supplies<br>Other | $ | $ | $ |
| Liabilities from balance sheets | | | |
| Accounts payable . . . . . . . . .<br>Interest expense<br>Other . . . . . . . . . . . . . . . . . | $<br>(34a) | $<br>(34b) | $<br>(34) |

| | Total noncash expense adjustments (F) | $ | |
|---|---|---|---|
| | Net income from operations (C − D − E ± F = G) | $ | (28) |
| | Gain/loss on sale of capital assets (H) | $ | (52) |
| | Net business income (G ± H = I) | $ | (29) |
| | Nonbusiness income (J) | $ | (30) |
| | Net income (I + J = K) | $ | |

Transfer value [52] from income and expense summary and value [30] from SCF worksheet. Transfer totals [25], [26], [27], [28], [30], [31], and [34b] to financial worksheet and [29] to SOE.

*Source: The National Council for Agricultural Education, DECISIONS & DOLLARS, 1995

**FIGURE 5-9**

### Expenses.

Examples of cash operating expenses:

1. Seed
2. Fertilizer
3. Chemicals

Examples of noncash expenses:

1. Depreciation
2. Changes in (used in accrual accounting)
   a. unused supplies
   b. accounts payable
   c. accrued interest
   d. accrued taxes

**FIGURE 5-10**

### Classroom team activities enhance students' abilities to keep accurate records.

## ■ NET INCOME

Net income, or profit, is the income prior to paying income and social security taxes have been deducted. Retained earnings are the funds after deducting taxes. The object of the income statement is to calculate net income.

### NET INCOME FROM OPERATIONS

Regardless of which method of accounting is utilized, net income from operations is calculated by subtracting expenses and depreciation from revenues

(Figures 5–11 and 5–12). The differences between the two accounting methods include an adjustment for noncash expense and using gross versus cash revenue when using the accrual system (Figure 5–13).

## NET BUSINESS INCOME

Net business income (returns to capital, labor, and management) is calculated when the gain or loss of sale of capital assets is subtracted from net income from operations (Figures 5–11 and 5–12). Income and social security taxes are calculated and deducted from this figure, leaving net income as the final figure (Tables 5–9 and 5–10).

**FIGURE 5–11**

Income statement: Cash accounting.

```
      Cash revenues
    − Cash expenses
    = Net cash income


      Net cash income
    − Depreciation
    = Net income from operations


      Net income from operations
    ± Gain/loss on sale of capital assets
    = Net income
```

**FIGURE 5–12**

Income statement: Accrual accounting.

```
      Gross revenues
    − Cash expenses
    − Depreciation
    ± Noncash expense adjustment
    = Net income from operation


      Net income from operations
    ± Gain/loss on sale of capital assets
    = Net income
```

**FIGURE 5-13**

Students work together to complete an income statement on the school's greenhouse enterprise.

## Summary

The primary message in this chapter is that the cash and accrual accounting methods are important and, in fact, necessary for an accurate and complete financial picture of the business. Keeping records does not have to be complicated, but keeping appropriate and accurate records does have to be a priority.

### SUPERVISED EXPERIENCE

Landscaping is a growing occupational area across the country. These students are putting their classroom instruction to use with this school-based landscaping project. They have used their decision-making skills in submitting bids for jobs in the community. The students have also determined that an accurate understanding of the financial end of the business is just as important as understanding plant science concepts. Teams of students prepared comprehensive landscape plans for each job they were interested in. The team leaders were responsible for keeping an up-to-date list of landscape material prices. Others in the group promoted and marketed the team. All team members assisted with the actual landscaping and shared in the profits (Figure 5–14).

## TABLE 5–9   INCOME BY ENTERPRISE**

| | Date | Item | Code* | Total value | Personal value | ✓ if noncash |
|---|---|---|---|---|---|---|
| | | TOTAL: Brought forward from previous page | | | | |
| 1 | | | | | | |
| 2 | | | | | | |
| 3 | | | | | | |
| 4 | | | | | | |
| 5 | | | | | | |
| 6 | | | | | | |
| 7 | | | | | | |
| 8 | | | | | | |
| 9 | | | | | | |
| 10 | | | | | | |
| 11 | | | | | | |
| 12 | | | | | | |
| 13 | | | | | | |
| 14 | | | | | | |
| 15 | | | | | | |
| 16 | | | | | | |
| 17 | | | | | | |
| 18 | | | | | | |
| 19 | | | | | | |
| 20 | | | | | | |
| 21 | | | | | | |
| 22 | | | | | | |
| | | TOTAL: Carry forward to next page | | | | |

*Indicate category code from Table 5–6.
**Source: The National Council for Agricultural Education, DECISIONS & DOLLARS, 1995

(*continued*)

**TABLE 5–9    INCOME BY ENTERPRISE (*CONTINUED*)**

| | Enterprise | | | Enterprise | | | Enterprise | | |
|---|---|---|---|---|---|---|---|---|---|
| | No. | Quantity | Value | No. | Quantity | Value | No. | Quantity | Value |
| | TOTAL: Brought forward | | | TOTAL: Brought forward | | | TOTAL: Brought forward | | |
| 1 | | | | | | | | | |
| 2 | | | | | | | | | |
| 3 | | | | | | | | | |
| 4 | | | | | | | | | |
| 5 | | | | | | | | | |
| 6 | | | | | | | | | |
| 7 | | | | | | | | | |
| 8 | | | | | | | | | |
| 9 | | | | | | | | | |
| 10 | | | | | | | | | |
| 11 | | | | | | | | | |
| 12 | | | | | | | | | |
| 13 | | | | | | | | | |
| 14 | | | | | | | | | |
| 15 | | | | | | | | | |
| 16 | | | | | | | | | |
| 17 | | | | | | | | | |
| 18 | | | | | | | | | |
| 19 | | | | | | | | | |
| 20 | | | | | | | | | |
| 21 | | | | | | | | | |
| 22 | | | | | | | | | |
| | TOTAL: Carry forward | | | TOTAL: Carry forward | | | TOTAL: Carry forward | | |

**TABLE 5–10    EXPENSES BY ENTERPRISE\*\***

| | Date | Item | Code* | Total value | Personal value | ✓ if noncash |
|---|---|---|---|---|---|---|
| | | TOTAL: Brought forward from previous page | | | | |
| 1 | | | | | | |
| 2 | | | | | | |
| 3 | | | | | | |
| 4 | | | | | | |
| 5 | | | | | | |
| 6 | | | | | | |
| 7 | | | | | | |
| 8 | | | | | | |
| 9 | | | | | | |
| 10 | | | | | | |
| 11 | | | | | | |
| 12 | | | | | | |
| 13 | | | | | | |
| 14 | | | | | | |
| 15 | | | | | | |
| 16 | | | | | | |
| 17 | | | | | | |
| 18 | | | | | | |
| 19 | | | | | | |
| 20 | | | | | | |
| 21 | | | | | | |
| 22 | | | | | | |
| | | TOTAL: Carry forward to next page | | | | |

\*Indicate category code from Table 5–6.
\*\*Source: The National Council for Agricultural Education, Decisions & Dollars, 1995

(*continued*)

**TABLE 5–10    EXPENSES BY ENTERPRISE (*CONTINUED*)**

| | Enterprise | | | Enterprise | | | Enterprise | | |
|---|---|---|---|---|---|---|---|---|---|
| | No. | Quantity | Value | No. | Quantity | Value | No. | Quantity | Value |
| | TOTAL: Brought forward | | | TOTAL: Brought forward | | | TOTAL: Brought forward | | |
| 1 | | | | | | | | | |
| 2 | | | | | | | | | |
| 3 | | | | | | | | | |
| 4 | | | | | | | | | |
| 5 | | | | | | | | | |
| 6 | | | | | | | | | |
| 7 | | | | | | | | | |
| 8 | | | | | | | | | |
| 9 | | | | | | | | | |
| 10 | | | | | | | | | |
| 11 | | | | | | | | | |
| 12 | | | | | | | | | |
| 13 | | | | | | | | | |
| 14 | | | | | | | | | |
| 15 | | | | | | | | | |
| 16 | | | | | | | | | |
| 17 | | | | | | | | | |
| 18 | | | | | | | | | |
| 19 | | | | | | | | | |
| 20 | | | | | | | | | |
| 21 | | | | | | | | | |
| 22 | | | | | | | | | |
| | TOTAL: Carry forward | | | TOTAL: Carry forward | | | TOTAL: Carry forward | | |

**FIGURE 5-14**

These students are team leaders of the school-based landscape company.

## CAREER OVERVIEW

Students with a background in agriculture, science, and finances can look forward to the following careers:

High school diploma: Tree Surgeon Assistant; Feed Mill Equipment Operator

Associate degree: Agriculture Machinery Assembler; Artificial Breeding Technician

Bachelor's degree: Commodity Broker; Food Inspector

Advanced degree: Policy Analyst; Water Quality Specialist

## DISCUSSION QUESTIONS

1. Explain the importance of an income statement?
2. Describe the income statement's relationship with other base financial statements?
3. What is the structure of an income statement?
4. What are the major components of an income statement?
5. Identify three sources of revenue?
6. Identify three sources of expenses?
7. Explain cash accounting procedures?
8. Explain accrual accounting procedures?

9. Compare the cash and accrual methods of determining net income?

10. Distinguish between net business income and net income from operations?

## SUGGESTED ACTIVITIES

1. Keep track of financial transactions for a personal or school-related activity for a period of time and then categorize them as revenue and expense.

2. Invite a local business leader to explain how to prepare and use an income statement to analyze a business.

## GLOSSARY

**Accounts Payable**   Unpaid bills or debts.

**Accrual Accounting**   Income is reported in the year earned and expenses are deducted or capitalized in the year incurred. Method of accounting used to compute earnings to be reported to the Internal Revenue Service. Includes increases and decreases in inventory; annual income and expenses; and complete inventory. *See also* cash accounting.

**Accrued Interest**   Amount of interest that would be due if the note were paid off on the date the statement is prepared. Interest built up over time.

**Accrued Taxes**   Amount of tax liability that would be due if the taxes were paid on the date the statement is prepared. Taxes built up over time.

**Balance Sheet**   Financial statement of equity for an individual/business for a specific point in time. List of assets (current or non-current) and liabilities (current and non-current) of an individual or business. *See also* net worth statement.

$$\text{assets} - \text{liabilities} = \text{owner equity/net worth/net gain or loss}$$

**Cash Accounting**   Method of accounting in which income is credited to the year it is received and expenses are deducted in the year they are paid. *See also* accrual accounting.

**Expense**   Cost of goods or services involved in producing a product or service.

**Income**   Payment received for goods or services: can be cash or noncash.

**Income Statement**   A summary of revenues and expenses over a period of time.

**Net Income**   Revenue minus expenses. Calculated by matching revenues with the expenses incurred to create those revenues, plus the gain or loss on the normal sales of capital assets. *See also* profit.

**Profit**   The difference between income and expenditures; net income. *See also* net income.

**Revenue**   income to the business from the sale of goods or services or changes in inventory. It consists of the proceeds received or value created from current business operations.

**Statement of Cash Flows**   Arranges information into operating activities, investing activities, and financing activities. The sum of the net cash flows is the change in cash during an operating period.

**Statement of Owner Equity**  The firm's check on accuracy at a point in time. It verifies that the base financial statements are in agreement and depicts where changes in owner equity occur.

**Unused Supplies**  Supplies and materials used in production that were not consumed during the current year.

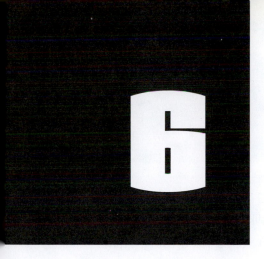

# Statement of Cash Flows

## OBJECTIVES

*After reading this chapter, you will be able to:*

- Explain the uses of a statement of cash flows.
- Describe the relationship of a statement of cash flows with other base financial statements.
- Identify the components of the statement of cash flows.
- Compare statement of cash flows and cash flow statement.
- Describe the advantages of using a statement of cash flows.
- Distinguish between statement of cash flows and income statement.
- Given a list of income and expenses, determine cash flow.

## KEY TERMS

| | | |
|---|---|---|
| Balance Sheet | Expenditure | Liquidity |
| Capital | GAAP | Receipt |
| Cash Flow Statement | Income | Revenue |
| Disbursement | Income Statement | Statement of Cash Flows |
| Equity | Investment | |

# BUSINESS PROFILE

*F*or all his life, Prunell dreamt of owning his own farm. His aptitude, living preferences, and likes and dislikes all make him suited to living in a rural area and producing crops for a living. Prunell knows that to begin farming and make an agribusiness profitable, he needs a large amount of money. He feels that he has enough money from an inheritance from his grandparents to purchase and begin the operation of a relatively small farm. He realizes that the first few years that he operates a farm will be tight financially, and he will have to manage his cash very carefully.

For several months, he searched for a farm for sale that would meet his expectations and be within his price range. He finally located a farm that seems promising. It is located in a good region and is currently in operation. The owner grows about 350 acres of soybeans, 160 acres of corn, and 200 acres of alfalfa. It is now late November, all the harvesting is completed, and the grain and alfalfa have all been sold.

The owner seems eager to sell and is willing to show Prunell all her farm records. Prunell knows that one of the best ways to tell how well any business is doing at a particular time is to examine the statement of cash flow. The owner seems to be a good record keeper and provides Prunell the statement for the past three months' financial transactions.

He is extremely satisfied with what he sees. This has been a good crop year. The 160 acres of corn yielded an average of 180 bushels per acre and brought $2.00 per bushel. The soybeans averaged 40 bushels per acre and brought $6.00 per bushel. A total gross income over the past three months of $141,600.00 was calculated. The expenses for that same period were for fuel, equipment repairs, maintenance, depreciation, labor, and transportation expenses. The total costs for the period were $67,144.57. The farm appears to have a profit of $74,455.43.

Prunell is well pleased with the current profit margin of the farm based on the statement of cash flows for the three months. However, there seems to be something that Prunell is overlooking that might have a tremendous impact on the profitability of the farm. Can you determine what Prunell is not considering? What would help him see the whole picture?

| Statement of Cash Flows for September through November | | |
|---|---|---|
| **Cash Income** | | |
| Sale of 28,800 bu. of corn @$2.00/bu. | $ 57,600.00 | |
| Sale of 14,000 bu. of soybeans @$6.00/bu. | $84,000.00 | |
| **Total Cash Received** | | **$141,600.00** |
| **Cash Paid for Operating Expenses** | | |
| Fuel | $ 4,826.54 | |
| Equipment repairs | $25,853.26 | |
| Machinery maintenance | $ 3,897.83 | |
| Transportation expenses | $ 1,487.98 | |
| Depreciation | $ 2,456.53 | |
| Labor | $28,622.43 | |
| **Total Expenses** | | **$ 67,144.57** |
| **Net Cash for the Period** | | **$ 74,455.43** |

**SOLUTION:** Prunell is not considering that much of the expenses for producing the crops were paid out earlier in the year. Seed, fertilizer, pesticides, fuel, equipment costs, labor, and other miscellaneous costs were incurred with the planting and cultivating of the crop. The income and expenses from the alfalfa were not included because that crop was harvested in the summer months. Also, the farmer might have gotten a higher price for the soybeans by holding the beans and selling them at a different time of the year. A statement of cash flows for the entire year would help him get a clearer overall picture of the profitability of the farm.

## Introduction

The purpose of this unit is to understand the concepts and components of the statement of cash flows and to use it in the financial management of a business. This is the third of the four important base financial statement chapters on the **balance sheet, income statement, statement of cash flows,** and reconciliation of owner equity. Information contained in financial statements is useful in the interpretation and understanding of the financial condition of a business enterprise.

The statement of cash flows (Figure 6–1) is a relatively new financial statement, not to be confused with a **cash flow statement** or budget. The Generally Accepted Accounting Principles **(GAAP)** require the inclusion of the statement of cash flows as one of four base financial statements.

## ■ STATEMENT OF CASH FLOWS VERSUS CASH FLOW STATEMENT

Both the statement of cash flows and cash flow statement, or budget, keep a summary of cash receipts and cash expenditures. The difference with the GAAP-accepted statement of cash flows is that the information is categorized as operating, investing, and financing activities. The statement of cash flows is an analysis and planning tool, whereas the cash flow statement is a projection tool (Tables 6–1 and 6–2).

> "*W*hat is the difference between an analysis and planning tool and a projection tool?" Megan asked.
>
> *M*s. Green stated, "The statement of cash flows, because it is directly tied to the chart of accounts that we learned about in Chapter 5, is organized to supply critical information in the business analysis process that you will learn about in Chapter 8. This analysis leads to business planning and decision making. The cash flow statement is a useful form that projects or records cash needs (inflows and outflows). Weekly, monthly, and/or quarterly operations are reliant on a consistent cash flow so that the payroll can be met and that short-term debts are paid."

**FIGURE 6–1**

Statement of cash flows.

Rearranges cash into:

  a. operating activities
  b. investing activities
  c. financing activities

Explains changes in cash or cash equivalents for period

**TABLE 6–1    STATEMENT OF CASH FLOWS***

| Enterprise | For the period | actual ☐ projected ☐ |
|---|---|---|
| **Cash flows from operating activities** | | |
| | | $ |
| | Net cash provided by operating activities | $ |
| **Cash flows from investing activities** | | |
| | | $ |
| | Net cash provided by investing activities | $ |
| **Cash flows from financing activities** | | |
| | | $ |
| | Net cash provided by financing activities | $ |
| | Net increase (decrease) in cash and cash equivalents | $ |
| | Cash and cash equivalents at beginning of year | $ |
| | Cash and cash equivalents at end of year | $ |

*Source: The National Council for Agricultural Education, DECISIONS & DOLLARS, 1995

**TABLE 6–2  CASH FLOW STATEMENT\*\***

Enterprise                     For the period

actual ☐
projected ☐

| | Total* | Quarter 1 Jan–Mar | Quarter 2 Apr–Jun | Quarter 3 Jul–Sep | Quarter 4 Oct–Dec |
|---|---|---|---|---|---|
| Beginning cash balance | * | | | | |
| | | | | | |
| Total cash available | * | | | | |
| | | | | | |
| Total cash required | | | | | |
| Cash position before savings and borrowing | * | | | | |
| | | | | | |
| Ending cash balance | * | | | | |

*Amounts in the "Total" column for "Beginning cash balance," "Total cash available," "Cash position before savings and borrowing," and "Ending cash balance" do not equal the sum of the amounts in the four quarters because the ending cash balance for a previous period is carried forward to be the beginning cash balance for the next period. All other amounts in the "Total" column are the sum of the four quarters.

\*\*Source: The National Council for Agricultural Education, Decisions & Dollars, 1995

# ■ USES OF A STATEMENT OF CASH FLOWS

A firm that stays in business must have enough cash to meet current obligations. Understanding how the cash flow is generated is extremely important for business owners. The primary uses of a statement of cash flows include evaluation and planning because the statement answers several important questions: Where did the cash flows come from? How were dollars spent? Is the change in cash flow consistent with the balance sheets cash position? Evaluation categories are summary information and historical analyses. Planning categories consist of projecting cash inflows and outflows, measuring **liquidity,** showing the ability to meet financial obligations, and maintaining a line of credit (Figure 6–2).

## SUMMARY OF RECEIPTS AND DISBURSEMENTS

Summarizing cash **receipts** and cash **disbursements, or expenditures,** is a traditional operation among businesses, but categorizing the information into operating, investing, and financing activities is a rather new concept. The net cash flow from the three new categories is summed to account for the change in cash and cash equivalents during the period.

## EVALUATING PAST FINANCIAL PRACTICES ON A HISTORICAL BASIS

Comparing statements of cash flows over a period of years provides invaluable business information. Seasonal fluctuations can be anticipated. Many small businesses have periods when tremendous amounts of expenditures are incurred without offsetting **revenue.** Likewise, **income** is often received during certain times of the year when expenses are low. Understanding where and when revenue and expenditure fluctuations may occur assists business owners in financial planning. That is, managers can forecast when operating loans are needed and when they can be paid back.

**FIGURE 6–2**

How to use statement of cash flows.

Evaluation: Historical basis
Where cash originated
How it was used
Planning: Projection basis
Helpful in testing ideas/decisions before committing dollars

"*H*ow does the past affect the future?" asked Troy.

"*T*he future has cycles and fluctuations that we need to plan for, and the past can provide us with information to assist in our planning. In agriculture, understanding the past is especially important because of the seasonal fluctuations that flood the market with produce, livestock, or other goods. This flooding causes a high supply, which leads to lower prices and less cash for operations. Producers who keep accurate records can then alter their production so that they sell during an off cycle, or they can sell their products through the futures market. The key point is that financial history is extremely important for future planning," responded Ms. Green.

## PROJECTING CASH INFLOWS AND OUTFLOWS

The evaluation of a business leads to planning and projecting possible cash inflows and outflows. Past business financial information is often the best indicator of future revenue and expenditures. This historical information, combined with current pricing information, is crucial when projecting cash inflows and outflows (Figure 6–3). The statement of cash flows can be used to provide this type of information prior to committing dollars to a new enterprise.

## MEASURING LIQUIDITY

The statement of cash flows' primary focus is to ensure that the business is liquid. That is, can the business meet its current obligations without disrupting

**FIGURE 6-3**

Projecting cash inflows and outflows is critical in the nursery business.

normal ongoing operations? A firm's financial goal is to have enough cash available to cover transactions.

## ABILITY TO MEET FINANCIAL OBLIGATIONS

Long-term debt demands repayment, and the statement of cash flows offer answers to the question of whether a business has the capacity to repay its long-term commitments or possible commitments. Lenders appreciate the accuracy of the financial information found in a statement of cash.

## MAINTAINING A LINE OF CREDIT

Obtaining operating loans and business accounts for short-term financing relies on accurate financial information. The statement of cash flows provides creditors with the data required to determine a line of credit.

---

"*R*emember, there are many reasons for keeping accurate records. For example, the statement of cash flows can be the key to opening the door of a prospective creditor or business partner," Ms. Green added.

*T*roy commented, "If I am going to realize my goal of owning a horticulture business, I am going to have to learn how to fill out all of these financial forms for two reasons, for planning and for obtaining credit."

---

## ■ RELATIONSHIP WITH OTHER BASE FINANCIAL STATEMENTS

The statement of cash flows integrates the concepts of liquidity and profitability by consolidating data from the income statement and balance sheet. In addition, when the statement of owner equity is added, so that the full set of financial statements is used, then financial risk and projections can also be determined (Figure 6–4).

The balance sheet indicates the firm's financial position at a given point in time. Reconciling the changes in net worth between balance sheets is the role of the accrual income statement net income figure. The statement of cash flows explains the net changes in balance sheet cash and cash equivalents. Confirming the accuracy and consistency of a business's financial information is the purpose of including all three base financial statements. As stated in previous chapters, the statement of cash flows must be completed for the same accounting time period as the balance sheet and income statement. This coordination provides the firm with its financial performance and position picture. Reconciling owner equity in the statement of owner equity, the fourth base financial statement, is explained in the next chapter.

**FIGURE 6-4**

Feasibility, risk, and profitability are determined by calculating the liquidity, repayment capacity, solvency, profitability, and financial efficiency of the business.

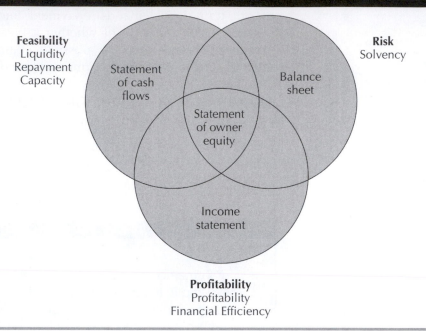

**Feasibility**
Liquidity
Repayment
Capacity

**Risk**
Solvency

Statement of cash flows

Balance sheet

Statement of owner equity

Income statement

**Profitability**
Profitability
Financial Efficiency

## ■ SOURCES OF CASH FLOW

Cash and cash equivalents are the only types of entries allowed on the statement of cash flows. Checking account balances, actual cash on hand, savings accounts, time certificates, government payment certificates, and the like are all included in this definition. Three areas are used to categorize statement of cash flows entries: operating, investing, and financing activities. This segregation allows for analysis within each area and provides the manager with critical information when making operating, investing, and financing decisions (Figure 6–5).

### OPERATING INCOME AND EXPENSES

The values in this section should correspond with the accrual income statement. They may consist of entries related to the production of goods and services, which includes consumable supplies. If the firm does not pay a wage or salary as a business expense for the labor and management contributions of family members, cash withdrawals for family or personal living and **investments** may also be a part of this section (Figure 6–6).

### INVESTMENTS

Understanding why purchases and sales of **capital** assets occur is answered in this section. It is important to know whether assets are being sold to account for operating cash deficiencies. Nonbusiness investments are included in this category. A negative cash flow requires either the operating or financing area to cover it, and a positive cash flow requires managers to scrutinize the records to determine if it is in the best interest of the business. (A positive cash flow can be generated by selling capital assets, but usually capital assets are required to generate income, so such a short-term solution to generate

**FIGURE 6-5**

Keeping accurate records is the first step in evaluating the business's past financial practices.

**FIGURE 6-6**

Operating activities: Statement of cash flows.

Cash flows resulting from the production of goods and services:

a. sale of product
b. purchase of operating inputs

cash flow can be disastrous for the business in the long term.) Examples of investments include land, buildings, machinery or equipment, personal loans to others, retirement accounts, and the like (Figure 6–7).

"*I*f I owned a greenhouse and added an automatic watering system, would it be considered a capital asset?" wondered Troy.

"*T*he watering system has a useful life greater than one year and, depending on the size of your greenhouse, is probably expensive enough to be included as a capital investment or asset," Ms. Green answered.

**FIGURE 6-7**

Investing activities: Statement of cash flows.

Cash flows resulting from purchases and sales of capital assets and nonbusiness investments such as:

a. land
b. buildings
c. machinery
d. equipment
e. loans to others

## FINANCIAL ACTIVITIES

Borrowing funds or other **equity** injections comprise the primary cash inflows. Loan repayments or capital lease repayments are the major cash outflow entries. This section provides the operator(s) with a summary of debt financing. It accounts for both operating and term debt, and this information is invaluable because it is instrumental in decisions concerning acquiring and repaying loans (Figure 6–8).

## ■ ADVANTAGES TO BUSINESS AND LENDERS

Operating in a businesslike manner is an important goal for many small businesses. Many such firms enter into the business world with a great idea or product, but little financial management background (Figure 6–9). Including a statement of cash flows along with the balance sheet and income statement when calculating financial status increases the completeness and accuracy of the information. This increase of accuracy is accomplished only if the financial entries are reliable.

The primary reason for improving financial information is to make management decisions with heightened confidence. The decision to obtain or pay off a

**FIGURE 6-8**

Financing activities: Statement of cash flows.

Cash flows resulting from borrowing or equity injection; outflows from loan repayment or capital lease repayment.

1. business borrowing
2. loan repayment
3. obtaining/disbursing equity funds
4. repaying principal portion of capital leases

**FIGURE 6-9**

Improving record keeping often takes the assistance of an expert.

loan can be reinforced with accurate and complete information (Figure 6–10). In some businesses, preparing a statement of cash flows for each enterprise is important to determine which areas are worth expanding and which areas need to be terminated.

"*M*y grandpa said that his dairy provides milk money every month and his soybeans give him a check only once a year. What would a statement of cash flows tell him?" queried Megan.

"*M*any small agribusinesses operate like your grandpa. That is, they have a product that supplies cash on a regular basis like milk, and they rely on it to pay the day-to-day bills. The large soybean check is probably used to pay operating and capital loans. A statement of cash flows would help your grandpa decide when would be the best time to sell his soybeans. Maybe he would find out that he should sell them in smaller quantities throughout the year," responded Ms. Green.

**FIGURE 6-10**

Advantages of good record keeping to business managers and lenders.

1. helps managers/lenders be more businesslike in financial analysis
2. improves information used in management/credit decisions

## ■ STATEMENT OF CASH FLOWS AND INCOME STATEMENT

Business operators need to have a clear understanding of the differences between the statement of cash flows and the income statement (Figure 6–11). Cash pays the bills and must be available when bills come due, or the business must borrow funds to meet its obligations. If the firm is unable to acquire funds, it may be in jeopardy of being terminated. This balance between cash inflows and outflows is constant throughout the year. Net income, on the other hand, is usually calculated annually and indicates overall profitability. Determining net cash from operating activities in the statement of cash flows and calculating net income from the accrual income statement should be done concurrently. This reconciliation from net income to net cash provides managers with the financial information that highlights the relationships among the base financial statements. In this case, it helps to distinguish between profitability and cash flow.

**FIGURE 6-11**

Distinguishing between liquidity, solvency, and profit is essential to understanding financial management.

## STATEMENT OF CASH FLOWS

The primary focus of the statement of cash flow is to track the actual cash flow during the past accounting period. The statement includes cash withdrawals, stock dividends, capital distribution, family living expenses, principal payments, the proceeds on new loans, the purchase or sale of capital items, cash from capital contributions or inheritances, and cash patronage dividends.

## INCOME STATEMENT

The income statement includes many noncash entries, and its purpose is to determine business profit or net income. Income statement entries that differ from the statement of cash flows are changes in inventories and accrued assets and liabilities, interest payments, depreciation expenses or gain/loss on sale of assets, and cash and noncash dividends.

## ■ COMPLETING THE STATEMENT OF CASH FLOWS

Consistent and accurate record keeping is essential when completing the statement of cash flows (Figure 6–12). Utilizing a coding system or chart of accounts is encouraged when entering ledger transactions. Coding income and expense entries enables business owners to conduct detailed financial analysis. Table 6–3 lists some common coding categories. Various coding systems are available, but the main categories are very similar. Differences occur within

**FIGURE 6-12**

The key to successful record keeping is entering data on a regular basis.

## TABLE 6–3    RECORD OF INCOME AND EXPENSES CODE SHEET*

The following items are coded and the code number should be included in the code* column on the following "Record on Income and Expenses" pages. You may include additional codes if you desire.

1 *cash withdrawals for family living:*
  1.1  cash withdrawal for family living
  1.2  cash withdrawals for investment into personal assets
2 *cash paid for operating activities:*
  2.1  cash paid for operating activities
    2.1.1  feeder animal
    2.1.2  feed purchases
    2.1.3  operating expenses—note you may want to further categorize this item such as:
      2.1.3.1—fuel; 2.1.3.2— oil/lubricants;
      2.1.3.3—fertilizer; and so on
    2.1.4  interest expenses—paid in cash or by loan renewal
      2.1.4.1  operating loans & accounts payable[33]
      2.1.4.2  term debt and capital leases[22]
    2.1.5  hedging account deposits
  2.2  cash expenses paid in nonbusiness operations
  2.3  income and social security taxes paid in cash
  2.4  extraordinary items paid in cash
3 *cash received from operating activities:*
  3.1  cash received from business operations (all cash sales from income statement)
    3.1.1  feeder livestock/poultry sales
    3.1.2  crops/feed
    3.1.3  custom work
    3.1.4  government payments
  3.2  cash received from nonbusiness income and operations
    3.2.1  wages
    3.2.2  interest
  3.3  extraordinary items received in cash
4 *cash paid for investing activities:*
  4.1  breeding and dairy livestock
  4.2  machinery and equipment

4.3  business real estate; other business assets
4.4  capital leased assets
4.5  bonds and securities; investments in other entities; other nonbusiness assets
5 *cash received from investing activities:*
  5.1  raised breeding and dairy livestock (not capitalized and not depreciated)
  5.2  purchased and raised breeding and dairy livestock (capitalized and depreciated)
  5.3  machinery and equipment
  5.4  business real estate; other business assets
  5.5  bonds and securities; investments in other entities; other nonbusiness assets
6 *distribution of dividends, capital, or gifts:*
  6.1  cash portion
  6.2  property portion
7 *cash received from equity contributions:*
  7.1  gifts and inheritances[41]
    7.1.1  cash portion[41a]
    7.1.2  property portion[41b]
  7.2  additions to paid-in capital including investments of personal assets into the business[42]
    7.2.1  cash portion[42a]
    7.2.2  property portion[42b]
8 *principal paid on term debt and capital leases:*
  8.1  term debt principal payments: scheduled payments
  8.2  term debt principal payments: unscheduled payments
  8.3  principal portion of payments on capital leases
9 *operating loans received (including interest paid by loan renewal):*
10 *proceeds from term debt (term debt financing; loans received):*
11 *operating debt principal payments:*

Note: It is important to record noncash items (such as: receiving pasture in exchange for labor, no cash was involved, but a value should be established on both pasture and labor). Include them as you would cash items, place a value on the items and identify them with a code. You may want to separate cash and noncash items during your analysis.
Note: Values[22], [33], [41], and [42] are calculated on Income and Expense Summary Schedule.
*Source: The National Council for Agricultural Education, DECISIONS & DOLLARS, 1995

the categories when businesses customize the subcategories to fit their needs. Codes 1 through 3 are the operating activities, codes 4 and 5 are the investing activities, and codes 6 through 11 are the financing activities.

## CODE 1: CASH WITHDRAWALS FOR FAMILY LIVING

Providing for a family, whether it is one person or an entire multigeneration group, is a main concern of businesses. This entry actually includes withdrawals both for family living and for personal investments, such as a retirement account.

# CODE 2: CASH PAID FOR OPERATING ACTIVITIES

These separately identified categories from the income and expenses code sheet are combined on the statement of cash flows. These items account for the cash outflows related to the business operations (Table 6–4).

## TABLE 6–4   INCOME AND EXPENSE SUMMARY*

**Enterprise** _____

Add all income and expense items with the same code and record on this page. Transfer the totals to the statement of cash flows worksheet and the statement of cash flows.

Note: For analysis purposes, the previous third and fourth level categories (e.g., 2.1.1, 2.1.4.1, and so on) have been combined to the larger categories listed below. The third and fourth level categories needed for financial analysis are calculated on the Income and Expense Summary Schedule.

1 *cash withdrawals for family living:* $1.1 + 1.2 = {}^{(35)}$
  1.1 cash withdrawal for family living . . . . . . . . . . . . . . . . . . . . . . . . . . . . . . . . . . . . $ _____ E
  1.2 cash withdrawals for investment into personal assets. . . . . . . . . . . . . . . . . . . . . . $ _____ E
2 *cash paid for operating activities:*
  2.1 cash paid for operating activities: feeder animal, feed purchases,
     operating and interest expenses, and hedging account deposits . . . . . . . . . . . . . $ _____ E
  2.2 cash expenses paid in nonbusiness operations . . . . . . . . . . . . . . . . . . . . . . . . . $ _____ E
  2.3 income and social security taxes paid in cash . . . . . . . . . . . . . . . . . . . . . . . . . $ _____ E
  2.4 extraordinary items paid in cash . . . . . . . . . . . . . . . . . . . . . . . . . . . . . . . . . . . $ _____ E
3 *cash received from operating activities:*
  3.1 cash received from business operations
     (all cash sales from income statement). . . . . . . . . . . . . . . . . . . . . . . . . . . . . . $ _____ E
  3.2 cash received from nonbusiness income and operations . . . . . . . . . . . . . . . . . . . $ _____ E
  3.3 extraordinary items received in cash . . . . . . . . . . . . . . . . . . . . . . . . . . . . . . . . . $ _____ R
4 *cash paid for investing activities:*
  4.1 breeding and dairy livestock . . . . . . . . . . . . . . . . . . . . . . . . . . . . . . . . . . . . . . $ _____ E
  4.2 machinery and equipment . . . . . . . . . . . . . . . . . . . . . . . . . . . . . . . . . . . . . . . . $ _____ E
  4.3 business real estate; other business assets. . . . . . . . . . . . . . . . . . . . . . . . . . . . $ _____ E
  4.4 capital leased assets . . . . . . . . . . . . . . . . . . . . . . . . . . . . . . . . . . . . . . . . . . . . $ _____ E
  4.5 bonds and securities; investments in other entities; other nonbusiness assets. . . . . $ _____ E
5 *cash received from investing activities:*
  5.1 raised breeding and dairy livestock (not capitalized and not depreciated). . . . . . . . $ _____ R
  5.2 purchased and raised breeding and dairy livestock
     (capitalized and depreciated) . . . . . . . . . . . . . . . . . . . . . . . . . . . . . . . . . . . . . $ _____ R
  5.3 machinery and equipment . . . . . . . . . . . . . . . . . . . . . . . . . . . . . . . . . . . . . . . . $ _____ R
  5.4 business real estate; other business assets. . . . . . . . . . . . . . . . . . . . . . . . . . . . $ _____ R
  5.5 bonds and securities; investments in other entities; other nonbusiness assets. . . . . $ _____ R
6 *distribution of dividends, capital, or gifts:*
  6.1 cash portion . . . . . . . . . . . . . . . . . . . . . . . . . . . . . . . . . . . . . . . . . . . . . . . . . . $ _____ R/E
  6.2 property portion. . . . . . . . . . . . . . . . . . . . . . . . . . . . . . . . . . . . . . . . . . . . . . . . $ _____ R/E
7 *cash received from equity contributions:*
  7.1 gifts and inheritances. . . . . . . . . . . . . . . . . . . . . . . . . . . . . . . . . . . . . . . . . . . . $ _____ R
  7.2 additions to paid-in capital including investments of personal assets into the business. . . . . . . $ _____ R
8 *principal paid on term debt and capital leases:*
  8.1 term debt principal payments: scheduled payments . . . . . . . . . . . . . . . . . . . . . . . $ _____ E
  8.2 term debt principal payments: unscheduled payments . . . . . . . . . . . . . . . . . . . . . $ _____ E
  8.3 principal portion of payments on capital leases. . . . . . . . . . . . . . . . . . . . . . . . . . $ _____ E
9 *operating loans received (including interest paid by loan renewal):* . . . . . . . . . . . . . {}^{(53)} $ _____ R
10 *proceeds from term debt (term debt financing; loans received):* . . . . . . . . . . . . . . . {}^{(45)} $ _____ R
11 *operating debt principal payments:* . . . . . . . . . . . . . . . . . . . . . . . . . . . . . . . . . . . . {}^{(54)} $ _____ E

E – indicates expenses; R – indicates revenue; R/E – indicates either expense or revenue.
Transfer values {}^{(35), (53), (45), and (54)} to statement of cash flows.
*Source: The National Council for Agricultural Education, Decisions & Dollars, 1995

## CODE 3: CASH RECEIVED FROM OPERATING ACTIVITIES

The gross cash inflows include all the cash sales as recorded on the income statement. All cash categories, including commodity certificates received from the government, are transferred directly from the income statement. Several of the entries are "cash portion only" items. In addition, this category includes nonbusiness related income and extraordinary items received as cash.

## CODE 4: CASH PAID FOR INVESTING ACTIVITIES

Acquisitions of property, equipment, and other production assets held to produce goods or services (not to be confused with inventory items) are recorded here. Nonbusiness assets, bonds, securities, and investments in other entities are included as cash paid for investing activities. Some of the items included are calculated in Table 6–5, Income and Expense Summary—Gain/Loss on Sale of Capital Assets.

## CODE 5: CASH RECEIVED FROM INVESTING ACTIVITIES

Dispersing property, equipment, and other production assets provides revenue entries to the statement of cash flows. Similar items are included in codes 4 and 5. They are categorized as cash received or cash paid depending on whether they were acquired or disposed.

"*A*ll of these codes are beginning to make sense. When a financial transaction occurs, you have to be able to categorize the entry with a code and then the analysis follows," Troy expressed.

"*I* am glad to see that the codes help you categorize your financial entries. Correct coding is essential for the rest of the financial analysis to occur," stated Ms. Green.

## CODE 6: DISTRIBUTION OF DIVIDENDS, CAPITAL, OR GIFTS

These values are usually found directly on corporation or partnership financial distribution forms. The complicated portion of this section occurs if you are the owner, operator, or manager of a business that is calculating and distributing dividends, capital, or gifts. An income and expense summary schedule (Table 6–6) is utilized to accurately compute the values needed for this section.

## CODE 7: CASH RECEIVED FROM EQUITY CONTRIBUTIONS

This section combines personal and business assets for a complete financial picture. The two categories are (1) gifts and inheritances and (2) investing personal assets into the business (Table 6–7).

**TABLE 6–5   INCOME AND EXPENSE SUMMARY: GAIN/LOSS ON SALE OF CAPITAL ASSETS\*\***

**Enterprise**

| Capital asset account adjustment for sales and transfers of raised breeding and dairy livestock (not fully capitalized and not depreciated) | | | |
|---|---|---|---|
| | beginning inventory | ending inventory | adjustment (+ or −) |
| value of raised breeding and dairy livestock                 use column e or g | −$ | +$ | ±$ |
| cash sales of raised breeding and dairy livestock .................................................................. | | + | (48) |
| net capital asset account adjustment (A) | | | ±$ |

| Capital gain (loss) on breeding and dairy livestock (fully capitalized & depreciated) | | | |
|---|---|---|---|
| Sales of: | net sales amount | adjusted cost/ basis\* | gain (loss) (+ or −) |
| purchased and raised breeding and dairy livestock ............... | +$          (49) | −$ | ±$ |
| death loss or other casualty loss......................... | +$ | −$ | ±$ |
| total gain or loss (B) | | | ±$ |

| Capital gain (loss) on machinery, real estate and other assets | | | |
|---|---|---|---|
| Sales of: | net sales amount | adjusted cost/ basis\* | gain (loss) (+ or −) |
| machinery and equipment ..................................................... | +$          (50) | −$ | ±$ |
| business real estate and other business assets ..................... | +$          (51) | −$ | ±$ |
| total gain or loss (C) | | | ±$ |
| Gain/loss on sale of capital assets (A + B + C = D) | | ±$ | (52) |

\*Usually the adjusted cost/basis is the most current value from column "e" or "g" from the inventory sheets.
Transfer values (48), (49), (50), and (51) from statement of cash flows worksheet.
Transfer value (52) to income statements.
\*\*Source: The National Council for Agricultural Education, Decisions & Dollars, 1995

# CODE 8: PRINCIPAL PAID ON TERM DEBT AND CAPITAL LEASES

This section is categorized by scheduled and unscheduled term debt principal payments. The difference between the two is determined by the date when the payment is made. An unscheduled payment is one that is not required and that helps to eliminate term debt earlier than anticipated. A scheduled payment is one that is required on a certain date or a penalty is assessed. The final category is the principal portion of payment on capital leases. Determining which portion of a capital lease is principal or interest is often confusing, and it sometimes calls for the help of the leasing agency.

## TABLE 6-6 INCOME AND EXPENSE SUMMARY: SCHEDULES*

### Enterprise        Interest Expense Schedule

| | | | | |
|---|---|---|---|---|
| interest paid in cash or by renewal on operating loans and accounts payable: <br> Total of code # 2.1.4.1 = (A) | | | | (33) |
| interest paid in cash or by renewal on term debt and capital leases: <br> Total of code # 2.1.4.2 = (B) | | | | (22) |
| Total interest paid in cash or by renewal (A + B = C)(C) | | | $ | |

| Transfer values for (32a), (16a), (32b), and (16b) from balance sheets | beginning of period | end of period | adjustments (+ or −) | |
|---|---|---|---|---|
| accrued interest on current liabilities: $[(32b) - (32a) = (32)]$ | (32a) | (32b) | ± | (32) |
| accrued interest on non-current liabilities: $[(16b) - (16a) = (16)]$ | (16a) | (16b) | ± | (16) |
| $[(32a) + (16a) = (34a)]$ and $[(32b) + (16b) = (34b)]$ | (34a) | (34b) | | |
| $[(34a) + (34b) = (34)]$ | | (34) | | |
| Total change in accrued interest payable $[(32) + (16) = D]$ | | | $ | |

Transfer values (34a), (34b), and (34) to income statement: accrual.
Transfer values (16), (22), and (34b) to financial worksheet.

### Gifts & Inheritance Schedule

| | |
|---|---|
| cash portion of gifts and inheritances: <br> Total of code # 7.1.1 = (A) | (41a) |
| property portion of gifts and inheritances: <br> Total of code # 7.1.2 = (B) | (41b) |
| Total gifts and inheritances (A + B = C)(C) | (41) |

### Additions to Paid-in Capital Schedule

| | |
|---|---|
| cash portion of additions to paid-in capital including investments of personal assets into the business: <br> Total of code # 7.2.1 = (A) | (42a) |
| property portion of additions to paid-in capital including investments of personal assets into the business: <br> Total of code # 7.2.2 = (B) | (42b) |
| Total additions to paid-in capital (A + B = C)(C) | (42) |

Transfer values (41a) and (42a) to statement of cash flows worksheet.
Transfer values (41) and (42) to statement of owner equity.
*Source: The National Council for Agricultural Education, DECISIONS & DOLLARS, 1995

## CODE 9: OPERATING LOANS RECEIVED

If the business is able to lend operating funds to others, it records any payments received, including interest, in this section. A completed statement of cash flows (Table 6–8) indicates where all the entries are recorded.

## TABLE 6-7  STATEMENT OF CASH FLOWS WORKSHEET*

**Enterprise**          **For the period**

### Cash flows from operating activities

2 *cash paid for operating activities:*
   2.1 cash paid for operating activities: feeder animal, feed purchases, operating and
        interest expenses, and hedging account deposits . . . . . . . . . . . . . . . . . . . . . . . . . . . . . . . . . . . . . . . . . (a)_____
   2.2 cash expenses paid in nonbusiness operations . . . . . . . . . . . . . . . . . . . . . . . . . . . . . . . . . . . . . . . . . (b)_____
   2.3 income and social security taxes paid in cash . . . . . . . . . . . . . . . . . . . . . . . . . . . . . . . . . . . . . . . (c)_____ [31]
   2.4 extraordinary items paid in cash . . . . . . . . . . . . . . . . . . . . . . . . . . . . . . . . . . . . . . . . . . . . . . . . . . . (d)_____
                 TOTAL cash paid for operating activities [a + b + c + d = (36)]:_____

3 *cash received from operating activities:*
   3.1 cash received from business operations (all cash sales from income statement) . . . . . . . . . . . . . . . . . (a)_____
   3.2 cash received from nonbusiness income and operations . . . . . . . . . . . . . . . . . . . . . . . . . . . . . . . . . . (b)_____ [30]
   3.3 extraordinary items received in cash . . . . . . . . . . . . . . . . . . . . . . . . . . . . . . . . . . . . . . . . . . . . . . . . (c)_____
               TOTAL cash received from operating activities [a + b + c = (37)]:_____

### Cash flows from investing activities

4 *cash paid for investing activities:*
   4.1 breeding and dairy livestock . . . . . . . . . . . . . . . . . . . . . . . . . . . . . . . . . . . . . . . . . . . . . . . . . . . . . . (a)_____
   4.2 machinery and equipment . . . . . . . . . . . . . . . . . . . . . . . . . . . . . . . . . . . . . . . . . . . . . . . . . . . . . . . . (b)_____
   4.3 business real estate; other business assets . . . . . . . . . . . . . . . . . . . . . . . . . . . . . . . . . . . . . . . . . . (c)_____
   4.4 capital leased assets . . . . . . . . . . . . . . . . . . . . . . . . . . . . . . . . . . . . . . . . . . . . . . . . . . . . . . . . . . . (d)_____
   4.5 bonds and securities; investments in other entities; other nonbusiness assets . . . . . . . . . . . . . . . . . (e)_____
                TOTAL cash paid to purchase [a + b + c + d + e = (38)]:_____

5 *cash received from investing activities:*
   5.1 raised breeding and dairy livestock (not capitalized and not depreciated) . . . . . . . . . . . . . . . . . . . . . (a)_____ [48]
   5.2 purchased and raised breeding and dairy livestock (capitalized and depreciated) . . . . . . . . . . . . . . . (b)_____ [49]
   5.3 machinery and equipment . . . . . . . . . . . . . . . . . . . . . . . . . . . . . . . . . . . . . . . . . . . . . . . . . . . . . . . . (c)_____ [50]
   5.4 business real estate; other business assets . . . . . . . . . . . . . . . . . . . . . . . . . . . . . . . . . . . . . . . . . . (d)_____ [51]
   5.5 bonds and securities; investments in other entities; other nonbusiness assets . . . . . . . . . . . . . . . . . (e)_____
            TOTAL cash received from investing activities [a + b + c + d + e = (39)]:_____

### Cash flows from financing activities

6 *distribution of dividends, capital or gifts:*
   6.1 cash portion . . . . . . . . . . . . . . . . . . . . . . . . . . . . . . . . . . . . . . . . . . . . . . . . . . . . . . . . . . . . . . . . . (a)_____ [40a]
   6.2 property portion . . . . . . . . . . . . . . . . . . . . . . . . . . . . . . . . . . . . . . . . . . . . . . . . . . . . . . . . . . . . . . (b)_____ [40b]
             TOTAL distribution of dividends, capital or gifts [a + b = (40)]:_____

7 *cash received from equity contributions:*
   7.1 cash portion of gifts and inheritances [41a] . . . . . . . . . . . . . . . . . . . . . . . . . . . . . . . . . . . . . . . . . . (a)_____
   7.2 cash portion of additions to paid-in capital including investments of personal assets
        into the business [42a] . . . . . . . . . . . . . . . . . . . . . . . . . . . . . . . . . . . . . . . . . . . . . . . . . . . . . . . . . (b)_____
          TOTAL (cash portion) received from equity contributions [a + b = (43)]:_____

8 *principal paid on term debt and capital leases:*
   8.1 term debt principal payments: scheduled payments . . . . . . . . . . . . . . . . . . . . . . . . . . . . . . . . . . . . . (a)_____
   8.2 term debt principal payments: unscheduled payments . . . . . . . . . . . . . . . . . . . . . . . . . . . . . . . . . . (b)_____
   8.3 principal portion of payments on capital leases . . . . . . . . . . . . . . . . . . . . . . . . . . . . . . . . . . . . . . . (c)_____
          TOTAL principal payments on term debt and capital leases [a + b + c = (44)]:_____
       TOTAL annual scheduled principal payments on term debt and capital leases [a + c = (23)]:_____

Transfer values [41a] and [42a] from income and expense summary.
Transfer values [48], [49], [50], and [51] to income and expense summary; value [30] to income statements; values [36], [37], [38], [39], [40a], [43], and [44] to the statement of cash flows; value [40] to statement of owner equity and values [23] and [31] to financial worksheet.
*Source: The National Council for Agricultural Education, DECISIONS & DOLLARS, 1995

## TABLE 6-8  COMPLETED STATEMENT OF CASH FLOWS*

| Enterprise | For the period | actual ☐ projected ☐ |
|---|---|---|
| **Cash flows from operating activities** | | |
| 1 cash withdrawals for family living: ............................................ | | (35) |
| 2 cash paid for operating activities: ............................................ | | (36) |
| 3 cash received from operating activities:........................................ | | (37) |
| other | | |
| | Net cash provided by operating activities (A) | $ |
| **Cash flows from investing activities** | | |
| 4 cash paid for investing activities:............................................ | | (38) |
| 5 cash received from investing activities:........................................ | | (39) |
| other | | |
| | Net cash provided by investing activities (B) | $ |
| **Cash flows from financing activities** | | |
| 6 distribution of dividends, capital, or gifts (cash portion): ...................... | | (40a) |
| 7 cash received from equity contributions: ...................................... | | (43) |
| 8 principal paid on term debt and capital leases: ................................ | | (44) |
| 9 operating loans received (including interest paid by loan renewal): .............. | | (53) |
| 10 proceeds from term debt (term debt financing; loans received): ................ | | (45) |
| 11 operating debt principal payments: .......................................... | | (54) |
| other | | |
| | Net cash provided by financing activities (C) | $ |
| | Net increase (decrease) in cash and cash equivalents (A + B + C = D) | $ |
| | Cash and cash equivalents at beginning of year (E) | $ (15) |
| | Cash and cash equivalents at end of year (D + E = F) | $ |

Transfer value (15) from beginning balance sheet statement.
Transfer values (35), (45), (53), and (54) from income and expenses summary.
Transfer values (36), (37), (38), (39), (40a), (43), and (44) from the statement of cash flows worksheet.
Transfer value (35) to statement of owner equity.
Transfer values (35), (36), (37), (38), (39), (40a), (43), (44), and (45) to financial worksheet.
*Source: The National Council for Agricultural Education, DECISIONS & DOLLARS, 1995

## CODE 10: PROCEEDS FROM TERM DEBT

Although unusual for most small businesses, this category is where payments received for term debt loans are summarized. Some firms specialize in making loans, but for most firms this category goes unused.

## CODE 11: OPERATING DEBT PRINCIPAL PAYMENTS

Different from code 8, this section records the principal payments for operating debt. For example, most small businesses require operating loans, and the principal portion of the repayment is entered in this section (Table 6–9).

### TABLE 6-9   COMPLETED CASH FLOW STATEMENT**

Enterprise                    For the period                    actual ☐  projected ☐

| | Total* | Quarter 1 Jan–Mar | Quarter 2 Apr–Jun | Quarter 3 Jul–Sep | Quarter 4 Oct–Dec |
|---|---|---|---|---|---|
| Beginning cash balance | * | | | | |
| | | | | | |
| Total cash available | * | | | | |
| | | | | | |
| Total cash required | | | | | |
| Cash position before savings and borrowing | * | | | | |
| | | | | | |
| Ending cash balance | * | | | | |

*Amounts in the "Total" column for "Beginning cash balance," "Total cash available," "Cash position before savings and borrowing," and "Ending cash balance" do not equal the sum of the amounts in the four quarters because the ending cash balance for a previous period is carried forward to be the beginning cash balance for the next period. All other amounts in the "Total" column are the sum of the four quarters.
**Source: The National Council for Agricultural Education, DECISIONS & DOLLARS, 1995

## OTHER

Because of the diversity of today's businesses, this category allows for any unaccounted cash flow entries.

## Summary

Any business needs to maintain a sufficient cash flow to meet current financial obligations without disrupting normal ongoing business operation. That is, the firm should strive for liquidity, at the same time remaining profitable for long-term viability. Balance sheets, income statements, and statements of cash flows are required financial statements when preparing a complete and accurate financial picture of the business. They are all necessary when preparing the statement of owner equity. The statement of cash flows is different from the cash flow statement, or budget, in that it categorizes cash inflows and outflows into three areas: operating, investing, and financing activities.

### SUPERVISED EXPERIENCE

Helping the environment is the goal of the students when they decided to assist the Game and Fish Department by restocking pheasants in their community. Increasing animal diversity with birds such as pheasants improves the environment and is a great learning experience (Figure 6–13). A small stipend is provided for feed and the eggs are furnished free, but the students have to develop a plan with a time line that corresponds to release dates established by the Game and Fish Department. The students are not planning

**FIGURE 6–13**

Increasing animal diversity by restocking pheasants in the community is the goal of this school-based supervised experience.

on making money on the project, but they do not plan to lose any either. Therefore, records are kept on all aspects of the project, and their teacher checks them regularly.

## CAREER OVERVIEW

One of the important lessons in this textbook is that understanding financial management is important to all people regardless of their career path. Of course, individuals who operate, manage, or own a business need greater entrepreneurial skills than those who work for a company or organization. Entrepreneurial skills include the record keeping and financial management skills taught in this book. However, improving personal financial understanding is also a goal of this book. All people will be faced with decisions regarding credit cards, obtaining loans for vehicles or homes, selecting retirement plans, choosing checking and savings accounts, purchasing appliances on credit, and so on. Understanding your personal financial position is a valuable skill that enhances your ability to optimize the use of your money. Experiences such as the students had in the Supervised Experience section provide career opportunities in many areas such as:

High school diploma: Custom Equipment Operator; Golf Course Technician

Associate degree: Agriculture Machinery Service Manager; Small Engine Mechanic

Bachelor's degree: Sales Representative; Grain Merchandiser

Advanced degree: Cooperative Extension Agent; Credit Analyst

## DISCUSSION QUESTIONS

1. Distinguish between a statement of cash flows and a cash flow statement.
2. Discuss the major uses of a statement of cash flows.
3. Identify and contrast the three activities that are categorized in a statement of cash flows.
4. Identify and illustrate several operating activity entries found in a statement of cash flows.
5. Identify and illustrate several investing activity entries found in a statement of cash flows.
6. Identify and illustrate several financing activity entries found in a statement of cash flows.
7. Explain the relationship of the statement of cash flows with the other base financial statements.
8. Contrast between a statement of cash flows and an income statement.
9. Explain the relationship between liquidity and the statement of cash flows.
10. Discuss profitability and its relationship with the statement of cash flows and the other base financial statements.

## SUGGESTED ACTIVITIES

1. Complete a statement of cash flows and cash flow statement for yourself or for a school-related entity such as a club, organization, activity, or the like.
2. Use the financial data from the previous chapters and prepare a statement of cash flows and a cash flow statement.
3. Research the historical development of the statement of cash flows.

## GLOSSARY

**Balance Sheet**   Financial statement of equity for an individual/business for a specific point in time. List of assets (current or non-current) and liabilities (current and non-current) of an individual or business. *See also* net worth statement.

$$\text{assets} - \text{liabilities} = \text{owner equity/net worth/net gain or loss}$$

**Capital**   Cash or liquid savings, machinery, livestock, buildings, and other assets that have a useful life of more than one year and that meet a specified minimum value.

**Cash Flow Statement**   Projected business payments and receipts associated without a particular business plan.

**Disbursement**   Funds paid out.

**Equity**   The net ownership of a business. It is the difference between the assets and liabilities of an individual or business as shown on the balance sheet or financial statement.

$$\text{assets} - \text{liabilities} = \text{owner equity}$$

**Expenditure**   Payment or expense incurred.

**GAAP**   Generally Accepted Accounting Principles; accounting guidelines used as standards in the financial world.

**Income**   Payment received for goods or services; can be cash or noncash.

**Income Statement**   A summary of revenues and expenses over a period of time.

**Investment**   The outlay of money, usually for income or profit.

**Liquidity**   Consisting of or capable of ready conversion to cash. Measures the ability of a business to meet financial obligations as they come due without disrupting normal ongoing operations.

**Receipt**   Document, usually a single sheet of paper, that acknowledges the receiving of goods, money, or services.

**Revenue**   Income to the business from the sale of goods or services or changes in inventory. It consists of the proceeds received or value created from current business operations.

**Statement of Cash Flows**   Arranges information into operating activities, investing activities, and financing activities. The sum of the net cash flows is the change in cash during an operating period.

# 7

# Statement of Owner Equity

## OBJECTIVES

*After reading this chapter, you will be able to:*

- Explain the purposes of the statement of owner equity.
- Describe its relationship with other base financial statements.
- Identify and discuss the structure and major components of a statement of owner equity.
- Given a sample balance sheet and a sample income statement, prepare a statement of owner equity.

## KEY TERMS

| | | |
|---|---|---|
| Additions to Paid-in Capital | Net Income | Statement of Cash Flows |
| Balance Sheet | Owner Equity | Statement of Owner Equity |
| Income Statement | Revenue | Stock |

# BUSINESS PROFILE

Amanda began a Supervised Agricultural Experience (SAE) program when she was in the ninth grade in high school. Her program consisted of raising rabbits for eight pet stores in the area. She began with only a pair of rabbits, and her operation grew to include 10 pairs of breeding rabbits and expanded to selling market rabbits to a local processing plant.

She is about to graduate from high school and wants to go to college. This requires that she move away from home, and she will not be able to continue the business. Her younger brother has entered the agricultural education program and wants to buy her business. Her brother cannot afford to pay her all of the money she has in the operation; so they have agreed to organize a partnership.

To make sure the partnership is designed fairly, a complete understanding of the business is necessary. Amanda has kept very good records and applies for both a proficiency award and her American FFA Degree. To help with organizing the records for the partnership application, she completes a statement of owner equity, which, when used as a financial tool, allows Amanda to determine exactly how the business has increased her financial worth. It begins with a statement of what her financial worth was at the beginning and end of each year. She is able to determine not only how much she is worth, but also where this worth is located (additions to paid-in capital, net income, gifts and inheritances, and so on). In addition, it assures the accuracy of her other records.

By having a good grasp of the amount of equity she has gained each year, Amanda is able to determine how much overall gain she has made in her rabbit producing business. This allows her to draw up an agreement with her brother that gives her a fair return on her investments in the business.

| Statement of Owner Equity for the Period | | |
|---|---|---|
| **Ending December 31, 19xx** | | |
| 1. Beginning owner equity | (A) | $2,689.34 |
| 2. Net income | (B) | ± $2,123.76 |
| 3. Gifts and inheritances | (C) | + $ 150.00 |
| 4. Additions to paid-in capital (investments of personal assets) | (D) | + $ 300.00 |
| 5. Withdrawals for personal uses | (F) | − $ 865.93 |
| 6. Total change in contributed capital | (B + C + D − F = G) | = $ 1,707.83 |
| 7. Ending owner equity | (A ± G = J) | = $ 4,397.17 |

## Introduction

The purpose of this chapter is to utilize a **balance sheet,** an **income statement,** and other important financial statements to reconcile **owner equity.** This is the last of four important base financial statements chapters, which define and explain the balance sheet, income statement, **statement of cash flows,** and **statement of owner equity.** Information contained in financial statements is useful in the interpretation and understanding of the financial condition of a business enterprise. Megan, Troy, and Ms. Green will seek the assistance of Mr. Wong, a loan officer, and Mrs. Ramirez, a bank official, as they attempt to reconcile questions on owner equity.

## ■ PURPOSES OF A STATEMENT OF OWNER EQUITY

The primary role of the statement of owner equity is to enable business operators to

1. conduct a final check on the accuracy of the change in owner equity between the beginning and ending balance sheets.
2. reconcile that change with **net income** and other possible changes in owner equity.

Owner equity is the most fundamental financial measure of the firm's financial position and performance. Verifying that the other base financial statements—balance sheet, income statement, and statement of cash flows—are in agreement is the goal of the statement of owner equity. The statement helps the business operator understand what occurred financially within the business during the financial year. It is a powerful addition to the base financial statements, and it offers incredible insight into the financial status of a business.

*T*roy states, "The idea of reconciliation seems similar to something we have previously discussed."

*M*s. Green comments, "Troy, you're very observant. A familiar reconciliation process is when you balance your checkbook account with the bank account statement. You are hoping that the accounts agree. If they don't agree, the discrepancies must be found and adjustments made. The statement of owner equity determines whether there is consistency among the base financial statements."

## ■ RELATIONSHIP WITH OTHER BASE FINANCIAL STATEMENTS

As previously mentioned, the statement of owner equity utilizes information from the other three base financial statements (Figure 7–1). The beginning and ending owner equity amounts come from the balance sheets. Net income

**FIGURE 7–1**

The statement of owner equity utilizes information from the other three base financial statements.

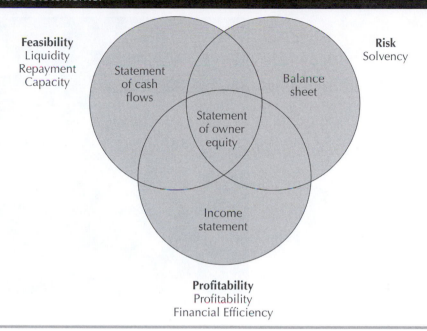

is calculated in the income statement. Other values are derived from income and expense summary schedules, the statement of cash flows worksheet, and the statement of cash flows. It all appears to be rather complicated, but like the explanation Ms. Green gave to Troy, the process is familiar and the figures should simply transfer if the records on the other statements have been kept up-to-date.

## ■ CHANGE IN TOTAL OWNER EQUITY

Even though the process is similar to balancing a checkbook, Megan and Troy understand that completing a statement of owner equity involves many more steps and financial statements. The following example illustrates the four primary areas where change in owner equity can occur (Figure 7–2).

### CHANGE IN RETAINED EARNINGS

For example, let's say that Troy began the year with $3,288 in owner equity (line 1 in Table 7–1 and entry 11 in Table 7–2). He earned $5,488 (entry A in Table 7–3) in gross cash **revenue** and ended the year with a net income of $1,376 (line 2 in Table 7–1 and entry I in Table 7–3), but he invested a portion of his income back into the business. His reinvestment of $842 is called "retained earnings" (line 7 in Table 7–1).

Troy's growth in owner equity of $842 from retained earnings was a 26% increase from the previous year. This practice of reinvesting income back into the business is common in most firms whose growth is a goal. In this case, Troy purchased a rototiller and some additional gardening tools. In fact, he

**FIGURE 7-2**

Collecting all the financial records for a school-based project involves teamwork and cooperation.

## TABLE 7-1   STATEMENT OF OWNER EQUITY*

Enterprise _____   For 12-month period ending _____

| | COST | MARKET VALUE |
|---|---|---|
| 1. Beginning owner equity[11] ..................... (A) | $_____ | $ 3,288.00 |
| 2. Net income (accrual)[29] ...................... (B) | ±_____ | ± 1,376.00 |
| 3. Gifts and inheritances[41] ..................... (C) | +_____ | + 500.00 |
| 4. Additions to paid-in capital including investments of personal assets into the business[42] ............ (D) | +_____ | + 150.00 |
| 5. Distributions of dividends, capital, or gifts made (cash or property)[40]......................... (E) | −_____ | − 0.00 |
| 6. Withdrawals for family living, gifts made, and investments into personal assets[35] ............ (F) | −_____ | − 1,184.00 |
| 7. Total change in contributed capital and retained earnings...............(B + C + D − E − F = G) | =_____ | = 842.00 |
| 8. Change in valuation equity[56 minus 55] ............. (H) | = NA | = (160.00) |
| 9. Ending owner equity ............. (A ± G ± H = J) | = $_____ | = $ 3,970.00 |
| Compare J with owner equity [14] on ending balance sheet. If different, explain discrepancies here: | | |

Transfer values [11] and [55] from beginning balance sheet.
Transfer value [56] from ending balance sheet.
Transfer values [41] and [42] from income and expense summary schedules.
Transfer value [29] from income statement.
Transfer value [40] from statement of cash flows worksheet.
Transfer value [35] from statement of cash flows.
*Source: The National Council for Agricultural Education, Decisions & Dollars, 1995

## TABLE 7-2 BEGINNING BALANCE SHEET*

**Balance Sheet Statement**

| Enterprise | | Beginning Values | | Date |
|---|---|---|---|---|

| Assets | Value | | Liabilities and Owner Equity | Value |
|---|---|---|---|---|
| Current assets | Cost | Market | Current liabilities | |
|   cash on-hand, checking, and savings | (15) | (15) |   accounts and notes payable | |
|   cash value—bonds, stocks, life insurance | | |   current portion of non-current debt[1] | |
|   notes and accounts receivable | | |   accrued interest on current liabilities | (32a) |
|   non-depreciable inventory | (1) | (1) |   accrued interest on non-current liabilities | (16a) |
|   other | | |   other | |
| Total current assets (A) | $ | $ | Total current liabilities (D) | $ |
| Non-current assets | Cost | Market | Non-current liabilities | Value |
|   non-depreciable inventory | (3) | (3) |   notes and chattel mortgages (total debt minus current portion)[1] | |
|   depreciable inventory | (5) | (5) |   real estate mortgages/contracts (total minus current portion) | |
|   real estate and improvements | (6) | (6) |   other | |
|   other | | | | |
| Total non-current assets (B) | $ | $ | Total non-current liabilities (E) | $ |

| | | | | Cost | Market |
|---|---|---|---|---|---|
| | | | Deferred taxes (T) | | $ |
| | | | Total liabilities (D + E + T = F) | $ (10) | $ (10) |
| | | | Owner equity: retained earnings contributed capital personal net worth valuation equity | | $ 200.00 (55) |
| | | | Total owner equity (C − F = G) | $ (11) | $3,288.00 (11) |
| Total assets (A + B = C)(C) | $ (9) | $ (9) | Total liabilities and owner equity (F + G = H) | $ | $ |

Transfer values (1), (3), (5), and (6) from inventories. Transfer value (11) to ending balance sheet and to statement of owner equity. Transfer values (16a) and (32a) to income & expense summary. Transfer value (15) to statement of cash flow and financial worksheet. Transfer values (9), (10), and (11) to financial worksheet. Transfer (T) from financial worksheet.

[1]The current portion of non-current debt when added to the notes and chattel mortgages (total minus current portion) should equal the total debt.

*Source: The National Council for Agricultural Education, DECISIONS & DOLLARS, 1995

## TABLE 7–3 INCOME STATEMENT: ACCRUAL*

Enterprise: _____     12-month period ending:_____

| Revenue | | | | Accrual |
|---|---|---|---|---|
| Cash revenue from ending balance of the record of income and expenses | | | | |
| | | | $ | |
| | | | Gross cash revenues (A) | $ 5,488.00 |

| Inventory adjustments | Inventories | | Difference (End. − Beg.) | |
|---|---|---|---|---|
| | Beg. | End. | | |
| | $ | $ | $ | |
| | | | Total inventory adjustment (B) | $ 522.00 |
| | | | Gross revenues (A ± B = C) | $ 6,010.00[25] |

| **Expenses** | | | | |
|---|---|---|---|---|
| Cash expenses from ending balance of the record of income and expenses | | | | |
| tax expense | | | $ [31] | |
| | | | Total cash expenses (D) | $ 4,166.00[26] |
| | | | Depreciation expense (E) | $ 344.00[27] |

**Noncash expense adjustments**

| Assets | Accounts | | Difference (End. − Beg.) | |
|---|---|---|---|---|
| | Beg. | End. | | |
| Unused supplies Other | $ | $ | $ | |

| Liabilities | | | | |
|---|---|---|---|---|
| Accounts payable..... Interest expense Other........................ | $ [34a] | $ [34b] | $ [34] | |

| | Accrual |
|---|---|
| Total noncash expense adjustments (F) | $ 86.00 |
| Net income from operations (C − D − E ± F = G) | $ 1,586.00[28] |
| Gain/loss on sale of capital assets (H) | $ 200.00[52] |
| Net business income (G ± H = I) | $ 1,386.00[29] |
| Nonbusiness income (J) | $ 300.00[30] |
| Net income (I + J = K) | $ 1,686.00 |

Transfer value [52] from income & expense summary & value [30] from SCF worksheet.
Transfer totals [25], [26], [27], [28], [30], [31] & [34b] to financial worksheet & [29] to SOE.
*Source: The National Council for Agricultural Education, DECISIONS & DOLLARS, 1995

**FIGURE 7-3**

Even a small landscaping maintenance business needs up-to-date record keeping for financial survival.

paid more than $842 for the new equipment, but, due to depreciation, the increased value is not the same as the purchase price (Figure 7–3).

## WITHDRAWALS

Troy noticed that, prior to determining his retained earnings, he had to deduct withdrawals for family living, gifts made, and investments in personal assets. His deductions of $1,184 were for personal living expenses. This seemed rather high, but a review of his figures indicated that the entry was correct (line 6 in Table 7–1 and entry 1 in Table 7–4).

*T*roy commented, "I remember spending more than $1,000 on new purchases, but I'm unsure why there are other deductions."

"*Y*ou are in luck. Mr. Wong, a loan officer from the credit union, is visiting our class today. Let's ask him," responded Ms. Green.

## CHANGE IN CONTRIBUTED CAPITAL

Mr. Wong examined Troy's financial records and quickly determined one source of the remaining revenue that was used to purchase the equipment. Like many growing businesses, Troy contributed funds or equity capital from nonbusiness sources. He received a $500 gift from his great-aunt, which he

## TABLE 7-4   STATEMENT OF CASH FLOWS*

| Enterprise | For the period | actual ☐ projected ☐ |
|---|---|---|
| **Cash flows from operating activities** | | |
| 1 cash withdrawals for family living: ........................................ | | 1,184.00(35) |
| 2 cash paid for operating activities: .......................................... | | (36) |
| 3 cash received from operating activities:................................... | | (37) |
| other | | |
| Net cash provided by operating activities (A) | | $ |
| **Cash flows from investing activities** | | |
| 4 cash paid for investing activities:.......................................... | | (38) |
| 5 cash received from investing activities:................................... | | (39) |
| other | | |
| Net cash provided by investing activities (B) | | $ |
| **Cash flows from financing activities** | | |
| 6 distribution of dividends, capital, or gifts (cash portion): ............ | | (40a) |
| 7 cash received from equity contributions:................................. | | (43) |
| 8 principal paid on term debt and capital leases: ....................... | | (44) |
| 9 operating loans received (including interest paid by loan renewal): ... | | (53) |
| 10 proceeds from term debt (term debt financing; loans received):....... | | (45) |
| 11 operating debt principal payments: ...................................... | | (54) |
| other | | |
| Net cash provided by financing activities (C) | | $ |
| Net increase (decrease) in cash and cash equivalents (A + B + C = D) | | $ |
| Cash and cash equivalents at beginning of year (E) | | $          (15) |
| Cash and cash equivalents at end of year (D + E = F) | | $ |

Transfer value (15) from beginning balance sheet statement.
Transfer values (35), (45), (53), and (54) from income and expenses summary.
Transfer values (36), (37), (38), (39), (40a), (43), and (44) from the statement of cash flows worksheet.
Transfer value (35) to statement of owner equity.
Transfer values (35), (36), (37), (38), (39), (40a), (43), (44), and (45) to financial worksheet.
*Source: The National Council for Agricultural Education, DECISIONS & DOLLARS, 1995

invested in the business (line 3 on Table 7–1 and entry 41 in Table 7–5). This reflects an increase in owner equity from a financial source outside the business (Figure 7–4).

Troy was still concerned because he remembered using some of his savings when he purchased the rototiller. In fact, Troy checked his records and informed Mr. Wong that he withdrew and used $150 from his savings. Mr. Wong told Troy that investments from personal accounts, **stocks,** or other

## TABEL 7–5   INCOME AND EXPENSE SUMMARY*

**Enterprise:**
**Interest Expense Schedule:**

| | beginning of period | end of period | adjustments (+ or −) | |
|---|---|---|---|---|
| interest paid in cash or by renewal on operating loans & accounts payable: Total of code # 2.1.4.1 = (A) | | | | (33) |
| interest paid in cash or by renewal on term debt & capital leases: Total of code # 2.1.4.2 = (B) | | | | (22) |
| Total interest paid in cash or by renewal (A + B = C)   $ | | | | |
| Transfer values for (32a), (16a), (32b), and (16b) from balance sheets: | beginning of period | end of period | adjustments (+ or −) | |
| accrued interest on current liabilities: $[(32b) - (32a) = (32)]$ | (32a) | (32b) | ± | (32) |
| accrued interest on non-current liabilities: $[(16b) - (16a) = (16)]$ | (16a) | (16b) | ± | (16) |
| $[(32a) + (16a) = (34a)]$ and $[(32b) + (16b) = (34b)]$ | (34a) | (34b) | | |
| $[(34a) + (34b) = (34)]$ | | (34) | | |
| Total change in accrued interest payable $[(32) + (16) = D]$   $ | | | | |

Transfer values (34a), (34b), and (34) to income statement: accrual.
Transfer values (16), (22), and (34b) to financial worksheet.

### Gifts & Inheritance Schedule:

| | |
|---|---|
| cash portion of gifts & inheritances: Total of code # 7.1.1 = (A) | 500.00 (41a) |
| property portion of gifts & inheritances: Total of code # 7.1.2 = (B) | 0.00 (41b) |
| Total gifts & inheritances (A + B = C) | 500.00 (41) |

### Additions to Paid-in Capital Schedule:

| | |
|---|---|
| cash portion of additions to paid-in capital including investments of personal assets into the business: Total of code # 7.2.1 = (A) | 150.00 (42a) |
| property portion of additions to paid-in capital including investments of personal assets into the business: Total of code # 7.2.2 = (B) | 0.00 (42b) |
| Total additions to paid-in capital (A + B = C) | 150.00 (42) |

Transfer values (41a), and (42a) to statement of cash flows worksheet.
Transfer values (41), and (42) to statement of owner equity.
*Source: The National Council for Agricultural Education, Decisions & Dollars, 1995

paid-in capital are classified as **additions to paid-in capital.** These contributions to the business are helpful, but their origin must be accurately noted in the business records. In this case, the $150 should be entered on line 4 on the statement of owner equity (Table 7–1 and entry 42 in Table 7–5). The two entries for the statement of owner equity total $650 and represent Troy's contributed capital to his business.

**FIGURE 7-4**

Daily worksheets enabled these students to understand their financial data entry and to make the reconciliation of owner equity a simple task.

Another contributed capital entry could be distributions of dividends, capital, or gifts made (cash or property). In Troy's case, there was no such entry (line 5 in Table 7–1 and entry 6 in Table 7–6).

## CHANGE IN ASSET VALUES

Mr. Wong's visit was over and most of the students were beginning to understand how to complete the statement of owner equity. Troy was reviewing his completed statement of owner equity when he noticed another entry that he did not understand, the change in valuation equity.

Mrs. Ramirez did, indeed, understand how to compute the change in valuation equity. Like all of the previous statement of owner equity entries, this entry

"*W*hat is the 'change in valuation equity' category on the statement of owner equity?" Troy queried.

"*I* am not sure," said Ms. Green. "My friend Mrs. Ramirez is a bank official and would know. Let's give her a call."

*M*rs. Ramirez said, "The $160 is the difference between the beginning-of-period valuation equity entry and the ending of period valuation equity entry."

**TABLE 7-6  STATEMENT OF CASH FLOWS WORKSHEET***

Enterprise _____    **For the period ending** _____

### Cash flows from operating activities

2 *cash paid for operating activities:*
2.1  cash paid for operating activities: feeder animal, feed purchases, operating and
  interest expenses, and hedging account deposits ........................................................... (a) _____
2.2  cash expenses paid in nonbusiness operations ........................................................... (b) _____
2.3  income and social security taxes paid in cash ........................................................... (c) _____ [31]
2.4  extraordinary items paid in cash .............................................................................. (d) _____
TOTAL cash paid for operating activities [a + b + c + d = (36)]: _____

3 *cash received from operating activities:*
3.1  cash received from business operations (all cash sales from income statement) .................. (a) _____
3.2  cash received from nonbusiness income and operations ............................................... (b) _____ [30]
3.3  extraordinary items received in cash ......................................................................... (c) _____
TOTAL cash received from operating activities [a + b + c = (37)]: _____

### Cash flows from investing activities

4 *cash paid for investing activities:*
4.1  breeding and dairy livestock ................................................................................... (a) _____
4.2  machinery and equipment ...................................................................................... (b) _____
4.3  business real estate; other business assets .............................................................. (c) _____
4.4  capital leased assets ............................................................................................ (d) _____
4.5  bonds and securities; investments in other entities; other nonbusiness assets ................... (e) _____
TOTAL cash paid to purchase [a + b + c + d + e = (38)]: _____

5 *cash received from investing activities:*
5.1  raised breeding and dairy livestock (not capitalized and not depreciated) ......................... (a) _____ [48]
5.2  purchased and raised breeding and dairy livestock (capitalized and depreciated) ............... (b) _____ [49]
5.3  machinery and equipment ...................................................................................... (c) _____ [50]
5.4  business real estate; other business assets .............................................................. (d) _____ [51]
5.5  bonds and securities; investments in other entities; other nonbusiness assets ................... (e) _____
TOTAL cash received from investing activities [a + b + c + d + e = (39)]: _____

### Cash flows from financing activities

6 *distribution of dividends, capital, or gifts:*
6.1  cash portion ........................................................................................................ (a) _____0.00 [40a]
6.2  property portion ................................................................................................... (b) _____0.00 [40b]
TOTAL distribution of dividends, capital, or gifts [a + b = (40)]: _____0.00

7 *cash received from equity contributions:*
7.1  cash portion of gifts and inheritances[41a] .................................................................. (a) _____
7.2  cash portion of additions to paid-in capital including
  investments of personal assets into the business[42a] ................................................... (b) _____
TOTAL (cash portion) received from equity contributions [a + b = (43)]: _____

8 *principal paid on term debt and capital leases:*
8.1  term debt principal payments: scheduled payments ..................................................... (a) _____
8.2  term debt principal payments: unscheduled payments .................................................. (b) _____
8.3  principal portion of payments on capital leases .......................................................... (c) _____
TOTAL principal payments on term debt and capital leases [a + b + c = (44)]: _____
TOTAL annual scheduled principal payments on term debt and capital leases [a + c = (23)]: _____

Transfer values [41a] and [42a] from income and expense summary. Transfer values [48], [49], [50], and [51] to income and expense summary; value [30] to income statements; values [36], [37], [38], [39], [40a], [43], and [44] to the statement of cash flows; value [40] to statement of owner equity; and values [23] and [31] to financial worksheet.
*Source: The National Council for Agricultural Education, Decisions & Dollars, 1995

is easy if the financial statements are accurate and up-to-date. She informed Troy that he was to subtract the beginning-of-period valuation equity entry (entry 55 in Table 7–2) from the end-of-period valuation equity entry (entry 56 in Table 7–7). The difference is entered on line 8 in the statement of owner equity, and in this case $160 is subtracted from the beginning owner equity.

**TABLE 7–7   ENDING BALANCE SHEET*[1]**

**Balance Sheet Statement**

Enterprise:                    Ending Values                    Date

| Assets | Value | | Liabilities and Owner Equity | Value | |
|---|---|---|---|---|---|
| Current assets: | Cost | Market | Current liabilities | | |
| cash on-hand, checking, & savings | | | accounts & notes payable | | |
| cash value—bonds, stocks, life ins. | | | current portion of non-current debt[2] | | |
| notes & accounts receivable | | | accrued interest on current liabilities | (32b) | |
| non-depreciable inventory | (2) | (2) | accrued int. on non-current liabilities | (16b) | |
| other | | | other | | |
| Total current assets (A) | $ (17) | $ (17) | Total current liabilities (D) | $ (18) | |
| Non-current assets | Cost | Market | Non-current liabilities | Value | |
| non-depreciable inventory | (4) | (4) | notes & chattel mortgages (total debt minus current portion)[2] | | |
| depreciable inventory | (7) | (7) | real estate mortgages/contracts (total minus current portion) | | |
| real estate & improvements | (8) | (8) | other | | |
| other | | | | | |
| Total non-current assets (B) | $ | $ | Total non-current liabilities (E) | $ | |
| | | | | Cost | Market |
| | | | Deferred taxes (T) | | $ |
| | | | Total liabilities (D + E + T = F) | $ (13) | $ (13) |
| | | | Owner equity: retained earnings contributed capital personal net worth valuation equity | | 40(56) |
| | | | Total owner equity (C – F = G) | $ (14) | $3970(14) |
| Total assets (A + B = C)(C) | $ (12) | $ (12) | Total liabilities and owner equity (F + G = H) | $ | $ |

Transfer values (2), (4), (7), and (8) from inventories. Transfer value (11) from beginning balance sheet. Transfer values (16b) and (32b) to income & expense summary. Transfer values (12), (13), (14), (17), and (18) to financial worksheet. Transfer (T) from financial worksheet.
[1]The current portion of non-current debt when added to the notes & chattel mortgages (total minus current portion) should equal the total debt.
*Source: The National Council for Agricultural Education, Decisions & Dollars, 1995

# ■ COMPARISON OF CALCULATED AND REPORTED OWNER EQUITY

The goal, when calculating owner equity, is to determine whether it matches the reported owner equity from the balance sheet. Usually, it matches, but when it doesn't an explanation is warranted. There are many other reasons for the reported and calculated owner equity figures to not match, but the reasons can commonly be found in several areas. First, personal withdrawals are sometimes higher than what was budgeted. Second, all additions or distribution of capital may not be reported. Third, the discrepancy comes from an unaccountable error in the income and expense ledger. The final major area of difference results by not recording all assets and liabilities on all balance sheets. Regardless of the reason, the reconciliation of owner equity calls for an explanation of all discrepancies to be included as footnotes to the statement of owner equity.

*M*egan stated, "Calculating these figures on an imaginary business is OK, but a personal business experience would be more meaningful."

*M*s. Green responded, "That is what the class is all about. Each student will develop a supervised experience program that involves keeping records. Some of the students will work on school-based projects, and others will develop their own personal projects."

## Summary

The growth of owner equity and understanding where the growth occurs are key elements in a successful business. Likewise, identifying areas that do not contribute to growth allows a business manager to make adjustments and keep the business profitable. The statement of owner equity is the form that assists the business manager to examine the changes in owner equity. The formal links to the other three base financial statements allow for a verification of the accuracy among all financial figures. The statement of owner equity, as it reconciles the base financial statements, forces them to obtain agreement and be consistent.

### *SUPERVISED EXPERIENCE*

Operating your own business requires determination and a plan. This young woman surveyed her neighborhood and determined that she could gross about $200 per week with a lawn care business. With equipment on loan from her parents, she marketed herself as reliable, friendly, and competitive (Figure 7–5). Eventually, she made enough money to buy her own equipment

**FIGURE 7–5**

This student has captured the neighborhood lawn care market through aggressive marketing and quality work.

and pay back her parents. Her lawn care business was so well established that she hired extra help in the summer. After purchasing new equipment, she invested her remaining profits in a college fund.

## CAREER OVERVIEW

Careers that require experience in agriculture, business, and science include:

High school diploma: Part-time Farmer; Greenhouse Grower
Associate degree: Data Entry Specialist; Livestock Buyer
Bachelor's degree: Wholesale Produce Manager; Regulatory Agent
Advanced degree: Geneticist; Waste Management Specialist

## DISCUSSION QUESTIONS

1. Explain the purpose of the statement of owner equity.
2. Describe the statement of owner equity's relationship with other base financial statements.
3. Identify the structure of a statement of owner equity.
4. Identify the major components of a statement of owner equity.
5. Describe the value of using contributed capital in a business.
6. Describe the shortcomings of using contributed capital in a business.
7. Describe the process of change in asset values.

## SUGGESTED ACTIVITIES

1. Given a balance sheet and an income statement from a personal or school-related activity, prepare a statement of owner equity.
2. Invite the school's business teacher to class to discuss the common discrepancies between what is reported as owner equity and what is calculated as owner equity.

## GLOSSARY

**Additions to Paid-in Capital**   Investments from personal accounts, stocks, or other nonbusiness sources.

**Balance Sheet**   Financial statement of equity for an individual/business for a specific point in time. List of assets (current or non-current) and liabilities (current and non-current) of an individual or business. *See also* net worth statement.

assets − liabilities = owner equity/net worth/net gain or loss

**Income Statement**   A summary of revenues and expenses over a period of time.

**Net Income**   Revenue minus expenses. Calculated by matching revenues with the expenses incurred to create those revenues, plus the gain or loss on the normal sales of capital assets. *See also* profit.

**Owner Equity**   Assets minus liabilities.

**Revenue**   Income to the business from the sale of goods or services or changes in inventory. It consists of the proceeds received or value created from current business operations.

**Statement of Cash Flows**   Arranges information into operating activities, investing activities, and financing activities. The sum of the net cash flows is the change in cash during an operating period.

**Statement of Owner Equity**   The firm's check on accuracy at a point in time. It verifies that the base financial statements are in agreement and depicts where changes in owner equity occur.

**Stock**   Transferable certificates representing partial ownership in a corporation.

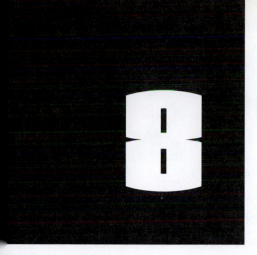

# Analyzing Financial Performance

## OBJECTIVES

*After reading this chapter, you will be able to:*

- Identify the five key areas examined during a complete financial analysis of a business.
- Recognize the 16 main measures that are used to evaluate the five key financial areas of a business.
- Explain the 16 main measures that are used to evaluate the five key financial areas of a business.
- Recognize guidelines for applying the use of financial measures.

# BUSINESS PROFILE

Jennifer inherited a business that sells farm chemicals to producers throughout a large portion of her state. The company sells a broad variety of chemicals, which include insecticides, herbicides, disinfectants, animal pharmaceuticals, fumigants, and other chemicals needed by agricultural producers. The business has been in the family for over 30 years, and she wants to be as successful as her father in supplying the chemical needs of producers who strive to earn a profit and protect the environment. The company operates in a wide area and has branch offices in five different counties that report back to the main office in Jennifer's hometown.

Before her father retired, Jennifer worked in the main office processing orders. Although she became an expert at her job, she lacked an understanding of the financial aspects of the business. As the owner, she wants to consider several changes. She thinks that the company could sell more chemicals with increased advertising. Also, the furniture and decor of the offices and stores are outdated. Jennifer wants to buy new furniture and completely redecorate all the offices and stores. She researched the costs of both the advertising and redecorating and prepared a budget. She realizes that this will take quite a lot of money from the operating budget of the company. She really wants to do the advertising and redecorating, but she also wants to make sure the company can comfortably pay all of its bills and still be profitable.

Mr. Ling has been the company's accountant for the past 18 years and understands the company's financial status. Jennifer requested a financial report from Mr. Ling indicating the financial performance of the company. He told her that he could send reports in any or all of the following areas:

liquidity

solvency

profitability

repayment capacity

financial efficiency

Which of the reports will provide Jennifer with information indicating whether she can increase advertising and begin redecorating?

**SOLUTION:** She needs a report of liquidity because this is a measure of the ability of a business to meet financial obligations as they come due without disrupting normal ongoing operations.

## Introduction

The purpose of this unit is to understand how to analyze the financial condition and performance of a business by utilizing its financial statements. The main reasons for keeping financial records are to accurately assess the status of a business and to assist in planning its future ventures. Conducting regular financial checkups and taking timely action based on adequate and accurate information improves financial decision making.

> *T*roy and Megan glanced ahead in this chapter and are concerned. They said to Ms. Green, "We see nothing but formulas, formulas, and more formulas."
>
> *M*s. Green remembered when she was first studying this subject as a student and how she had the same reaction. She also remembered learning, after years of experience, that applying these formulas to daily situations yielded valuable information about businesses.

## ■ FIVE KEY AREAS OF FINANCIAL ANALYSIS

Figure 8–1 presents general descriptions of the five areas that businesses (and individuals like Troy and Megan) evaluate when making financial decisions.

1. liquidity
2. solvency
3. profitability
4. repayment capacity
5. financial efficiency

**FIGURE 8–1**

Five areas of financial analyses.

**Liquidity**
Ability to meet obligations when due without disrupting normal operations.

**Solvency**
Borrowed capital in relation to owner equity capital invested in the business.

**Profitability**
Amount of profit from use of labor, management, and capital.

**Repayment Capacity**
Ability to repay debts from both business and nonbusiness income.

**Financial Efficiency**
Efficiency in use of labor, management, and capital.

*A*fter much careful thought, Ms. Green decided to help Troy and Megan see that they are already using informal versions of these formulas to make decisions every day. Ms. Green asked, "How do you evaluate whether you can afford a ticket to the new movie in town, 'Phantom-Shopper'?"

*T*roy said, "I look at how much cash I have today, and I compare it to how much money I need to spend between now and the weekend on things like lunch and snack food."

*M*s. Green responded, "You were evaluating your liquidity. Your cash is your current assets, and your commitments to buy lunch and other things by the weekend are like your current liabilities."

Sixteen main measures are used to evaluate the five areas of financial analysis (Table 8–1). The formulas used in calculating the 16 financial measures are listed in Table 8–2. An explanation of them follows.

## ■ LIQUIDITY

### 1. WORKING CAPITAL

Working capital indicates the extent to which current assets would cover current liabilities. It is expressed as a dollar value, and the higher the value is, the greater the **liquidity** is. However, the amount of working capital that is considered adequate is dependent on the size and needs of the business.

*M*egan remembered, "Some of these measures we learned about in Chapter 4 when we studied balance sheets."

*M*s. Green commented, "That is correct, you studied measures 1, 2, 4, 5, and 6."

### 2. CURRENT RATIO

Current ratio indicates the extent to which current assets would cover current liabilities if liquidated. (A ratio is one number divided by another to express a relationship.) However, unlike working capital, the current ratio is usually expressed as XX:1. If the ratio is greater than one to one (1:1), the business is considered liquid. If the ratio is less than 1:1, the business is considered not liquid, reflecting a high degree of short-term cash flow risks. The higher the ratio is, the more liquid the business is. Unlike working capital, this ratio is not dependent on the size of a business. Therefore, comparisons to other businesses can be made regardless of size differences (Figure 8–2).

## TABLE 8-1   SIXTEEN MAIN FINANCIAL MEASURES*

| Liquidity |
| --- |
| 1. Working capital |
| 2. Current ratio |
| 3. Cash flow coverage ratio |

| Solvency |
| --- |
| 4. Debt/asset ratio |
| 5. Equity/asset ratio |
| 6. Debt/equity ratio |

| Profitability |
| --- |
| 7. Return on assets |
| 8. Return on equity |
| 9. Operating profit margin ratio |

| Repayment Capacity |
| --- |
| 10. Term debt and capital lease coverage ratio |
| 11. Capital replacement and term debt repayment margin |

| Financial Efficiency |
| --- |
| 12. Asset turnover ratio |
| 13. Operating expenses ratio |
| 14. Depreciation expense ratio |
| 15. Interest expense ratio |
| 16. Net income from operations ratio |

*Source: The National Council for Agricultural Education, DECISIONS & DOLLARS, 1995

## TABLE 8-2   FINANCIAL RATIOS*

| Liquidity | | |
| --- | --- | --- |
| 1. Working capital | total current assets .................................................. <br> − *total current liabilities* ........................................... − | _____ <br> working capital = $ |
| 2. Current ratio | *total current assets* ................................................. <br> ÷ total current liabilities ............................................ ÷ | _____ <br> current ratio = |

*Source: The National Council for Agricultural Education, DECISIONS & DOLLARS, 1995

**TABLE 8-2    FINANCIAL RATIOS (*CONTINUED*)**

| | |
|---|---|
| 3. Cash flow coverage ratio | beginning cash<br>+ cash received from operating activities . . . . . . . . . . . . . . . . . . . . . . . . . . . . . . +<br>+ cash received from investing activities . . . . . . . . . . . . . . . . . . . . . . . . . . . . . +<br>+ proceeds from term debt. . . . . . . . . . . . . . . . . . . . . . . . . . . . . . . . . +<br>+ *cash received from equity contributions* . . . . . . . . . . . . . . . . . . . . +<br>÷ cash paid for operating activities. . . . . . . . . . . . . . . . . . . . . . . . . . . . ÷<br>+ cash paid for investing activities . . . . . . . . . . . . . . . . . . . . . . . . . . . . . +<br>+ principal paid on term loans and capital leases. . . . . . . . . . . . . . . . . . . +<br>+ cash portion distribution of dividends, capital, or gifts . . . . . . . . . . . . . . . . +<br><div align="right">cash flow coverage ratio =</div> |

### Solvency

| | |
|---|---|
| 4. Debt/asset ratio | *total liabilities*. . . . . . . . . . . . . . . . . . . . . . . . . . . . . . . . . . . . .<br>÷ total assets . . . . . . . . . . . . . . . . . . . . . . . . . . . . . . . . . . . . . . . . ÷<br><div align="right">debt/asset ratio =</div> |
| 5. Equity/asset ratio | *total equity*. . . . . . . . . . . . . . . . . . . . . . . . . . . . . . . . . . . . . .<br>÷ total assets . . . . . . . . . . . . . . . . . . . . . . . . . . . . . . . . . . . . . . . ÷<br><div align="right">equity/asset ratio =</div> |
| 6. Debt/equity ratio | *total liabilities*. . . . . . . . . . . . . . . . . . . . . . . . . . . . . . . . . . . . .<br>÷ total equity . . . . . . . . . . . . . . . . . . . . . . . . . . . . . . . . . . . . . . . ÷<br><div align="right">debt/equity ratio =</div> |

### Profitability

| | |
|---|---|
| 7. Return on assets | net income from operation . . . . . . . . . . . . . . . . . . . . . . . . . . . . . . . . . .<br>+ interest expense . . . . . . . . . . . . . . . . . . . . . . . . . . . . . . . . . . . . . . +<br>− *value of unpaid operator and family labor and management* . . . . . . . . . . . . −<br>÷ average total assets. . . . . . . . . . . . . . . . . . . . . . . . . . . . . . . . . . . ÷<br><div align="right">return on assets =</div> |
| 8. Return on equity | net income from operation . . . . . . . . . . . . . . . . . . . . . . . . . . . . . . . . . .<br>− *value of unpaid operator and family labor and management* . . . . . . . . . . . . −<br>÷ average total equity. . . . . . . . . . . . . . . . . . . . . . . . . . . . . . . . . . . ÷<br><div align="right">return on equity =</div> |

### Repayment Capacity

| | |
|---|---|
| 9. Operating profit margin ratio | net income from operation . . . . . . . . . . . . . . . . . . . . . . . . . . . . . . . . . .<br>+ interest expense . . . . . . . . . . . . . . . . . . . . . . . . . . . . . . . . . . . . . +<br>− *value of unpaid operator and family labor and management* . . . . . . . . . . . . −<br>÷ gross revenue . . . . . . . . . . . . . . . . . . . . . . . . . . . . . . . . . . . . . . ÷<br><div align="right">operating profit margin ratio =</div> |
| 10. Term debt and capital lease coverage ratio | net income from operation . . . . . . . . . . . . . . . . . . . . . . . . . . . . . . . . . .<br>+ total nonbusiness related income . . . . . . . . . . . . . . . . . . . . . . . . . . . . +<br>+ depreciation/amortization expense . . . . . . . . . . . . . . . . . . . . . . . . . . . +<br>+ interest on: term debt and capital leases. . . . . . . . . . . . . . . . . . . . . . . . +<br>− total income tax expense . . . . . . . . . . . . . . . . . . . . . . . . . . . . . . . . . −<br>− *withdrawals for family living*. . . . . . . . . . . . . . . . . . . . . . . . . . . . . . −<br>÷ annual scheduled principal and interest payments on term<br>    debt and on capital leases . . . . . . . . . . . . . . . . . . . . . . . . . . . . . . . ÷<br><div align="right">term debt and capital lease coverage ratio =</div> |

*(continued)*

**TABLE 8-2 FINANCIAL RATIOS (CONTINUED)**

| | |
|---|---|
| 11. Capital replacement and term debt repayment margin | net income from operation . . . . . . . . . . . . . . . . . . . . . . . . . . . . . . . . . . . .<br>+ total nonbusiness related income . . . . . . . . . . . . . . . . . . . . . . . . . . . . . . . .  +<br>+ depreciation/amortization expense . . . . . . . . . . . . . . . . . . . . . . . . . . . . .  +<br>− total income tax expense . . . . . . . . . . . . . . . . . . . . . . . . . . . . . . . . . . . .  −<br>− *withdrawals for family living* . . . . . . . . . . . . . . . . . . . . . . . . . . . . . . . .  −  _____<br>= capital replacement and term debt repayment capacity . . . . . . . . . . . . . .  =<br>− principal paid on term debt and capital leases . . . . . . . . . . . . . . . . . . . . .  −<br>capital replacement and term debt repayment margin = $ |
| **Financial Efficiency** ||
| 12. Asset turnover ratio | *gross revenue* . . . . . . . . . . . . . . . . . . . . . . . . . . . . . . . . . . . . . . . . . . . . . .  _____<br>÷ average total assets . . . . . . . . . . . . . . . . . . . . . . . . . . . . . . . . . . . . . . . . .  ÷<br>asset turnover ratio = |
| 13. Operating expenses ratio | total cash expense . . . . . . . . . . . . . . . . . . . . . . . . . . . . . . . . . . . . . . . . . .<br>− *depreciation expense* . . . . . . . . . . . . . . . . . . . . . . . . . . . . . . . . . . . . . . .  −  _____<br>÷ gross revenue . . . . . . . . . . . . . . . . . . . . . . . . . . . . . . . . . . . . . . . . . . . . .  ÷<br>operating expenses ratio = |
| 14. Depreciation expense ratio | *depreciation expense* . . . . . . . . . . . . . . . . . . . . . . . . . . . . . . . . . . . . . . . .  _____<br>÷ gross revenue . . . . . . . . . . . . . . . . . . . . . . . . . . . . . . . . . . . . . . . . . . . . .  ÷<br>depreciation expense ratio = |
| 15. Interest expense ratio | *total interest expense* . . . . . . . . . . . . . . . . . . . . . . . . . . . . . . . . . . . . . . .  _____<br>÷ gross revenue . . . . . . . . . . . . . . . . . . . . . . . . . . . . . . . . . . . . . . . . . . . . .  ÷<br>interest expense ratio = |
| 16. Net income from operations ratio | *net income from operations* . . . . . . . . . . . . . . . . . . . . . . . . . . . . . . . . . . .  _____<br>÷ gross revenue . . . . . . . . . . . . . . . . . . . . . . . . . . . . . . . . . . . . . . . . . . . . .  ÷<br>net income from operations ratio = |

## 3. CASH FLOW COVERAGE RATIO

This ratio assesses cash flow by looking more closely at the major categories that affect cash flow. Like the current ratio, the higher the ratio is, the greater the liquidity is. The primary limitation is that it tells only about past cash flows. It has

**FIGURE 8-2**

It is important to take time to understand the 16 financial measures.

limited reliability and accuracy for the projection of cash inflows and outflows. An adequate margin is needed to cover variability in cash flow.

## ■ SOLVENCY

Whereas liquidity measures involve primarily current assets and current liabilities, **solvency** measures deal with total assets and total liabilities. Therefore, solvency measures are good indicators of the debt repayment capacity of a business if all assets are liquidated (Figure 8–3).

### 4. DEBT-TO-ASSET RATIO

The debt-to-asset ratio measures the proportion of total assets owed to creditors, as opposed to total assets owed to owners of the business. The financial position of the business is expressed as a proportion of the total assets owed to creditors. The higher the ratio is, the greater the risk is that **debt** owed to creditors cannot be repaid.

### 5. EQUITY-TO-ASSET RATIO

The equity-to-asset ratio measures the proportion of total assets financed by the business owners, as opposed to total assets owed to creditors. It measures the proportion of assets financed by owner equity capital. The lower the ratio is, the greater the risk is that the business debt cannot be repaid.

### 6. DEBT-TO-EQUITY RATIO

Like the debt-to-asset ratio, when this ratio is low, creditors have less money in the business than the owner. This ratio compares debt to equity instead of to assets. Therefore, it tells more about the proportion of debt capital to equity capital. The higher the ratio value is, the more total capital is supplied by the creditors and consequently less by the owner (Figure 8–4).

**FIGURE 8-3**

This student, like many others, has a small supervised experience program. She will utilize only five of the financial measures when she analyzes her horticulture business.

**FIGURE 8-4**

Debt repayment capacity is especially important when returns on investment are not realized for several years, as was the case for this school orchard.

## ■ PROFITABILITY

When you are dealing with **profitability,** the focus shifts from the balance sheet (assets, liabilities, and owner equity) to the income statement (**revenues** and expenses). Businesses can increase profit by increasing the profit per unit produced or by increasing the volume of production (assuming the business venture is profitable). The three financial measures indicate the extent to which a business generates a profit or net income from the use of labor, management, and capital.

"*B*efore I enrolled in this class, I thought profit was the most important issue to businesses," Troy stated.

"*P*rofit is extremely important to all businesses, but a clear understanding of the five financial criteria—liquidity, solvency, profitability, repayment capacity, and financial efficiency—is also necessary for successful businesses," Ms. Green responded.

## 7. RETURN ON ASSETS

The return-on-assets ratio measures the rate of return on assets used in the business. The higher the ratio is, the more profitable is the business. This ratio is the most meaningful measure for comparing one business to another if the market value approach is used to value assets.

**FIGURE 8-5**

The profit earned on her customized equine grooming business allowed this young woman to purchase feed and supplies for her own horse.

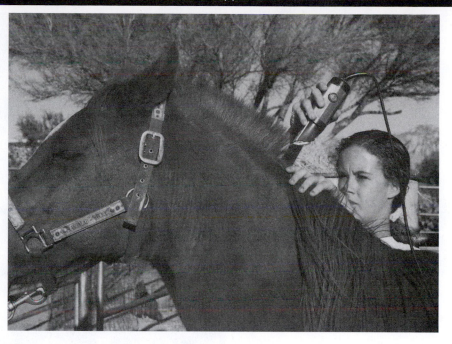

## 8. RETURN ON EQUITY

The return-on-equity ratio measures the rate of return on the owner's equity capital used in a business. The higher the ratio is, the more profitable is the business. This measure is not significant unless the market value approach is used to value assets (Figure 8–5).

## 9. OPERATING PROFIT MARGIN RATIO

This ratio measures the return to capital per dollar of gross revenue. It can be used to compare one business to another. Use accrual income measures for this measure to make trends meaningful. A high ratio indicates high profitability.

## ■ REPAYMENT CAPACITY

**Repayment capacity** measures the ability of a borrower to repay term business debt from business and nonbusiness income. Principal payments on term loans must come from net income after owner withdrawals, income taxes, and social security taxes (Figure 8–6).

## 10. TERM DEBT AND CAPITAL LEASE COVERAGE RATIO

This ratio measures the business's ability to pay term debt and capital lease payments. The more the ratio is over one to one (1:1), the greater the margin is to cover payments and the greater the flexibility is to withstand adversity.

**FIGURE 8-6**

Repayment capacity is a nonissue with this student, who leases equipment on a job-by-job basis for his landscape business.

## 11. CAPITAL REPLACEMENT AND TERM DEBT REPAYMENT MARGIN

This measure is actually a dollar value, not a ratio. Unlike the preceding ratio of term debt and capital lease coverage, this measure does not include interest on term debt and capital leases. So, if the computed margin is equal to interest on term debt and capital leases, then there is no room to withstand adversity. If the computed margin is greater than interest on term debt and capital leases, then there is room to get through financial hard times.

## ■ FINANCIAL EFFICIENCY

**Financial efficiency** analysis deals with the relationships between inputs and outputs. It measures the degree of efficiency in using labor, management, and capital. Gross revenue is the primary measure of outputs, and it is used in all of these measures.

> **"C**alculating financial efficiency measures looks simple when compared to the repayment capacity measures," Megan observed.
>
> **"T**hey are simple calculations and each formula includes gross revenue," Ms. Green added.

## 12. ASSET TURNOVER RATIO

The asset turnover ratio measures how effective assets are at generating revenue. The higher the ratio is, the more efficiently assets are being used to generate revenue.

The next four operating ratios represent the total composition of gross revenue. In percentage terms, they reflect the allocation of 100% of the gross revenues of a business.

## 13. OPERATING EXPENSES RATIO

Operating expenses are compared to gross revenue.

## 14. DEPRECIATION EXPENSE RATIO

Depreciation expense is compared to gross revenue, and the ratio varies by business type.

## 15. INTEREST EXPENSE RATIO

Interest expense is compared to gross revenue. The sum of ratios 13 through 15 expresses total business expenses per dollar of gross revenue.

## 16. NET INCOME FROM OPERATIONS RATIO

Net income from operations is compared to gross revenue and is calculated on a pretax basis.

> **M**s. Green knows that there is a lot of information in this chapter and agreed that everyone should reflect on what was just read.
>
> **"N**ow that we have the financial measures, what do we do with them?" Megan and Troy pondered.
>
> **"T**he last section of this chapter contains some guidelines that outline how to apply these financial measures," Ms. Green said.

## ■ GUIDELINES FOR APPLYING THE USE OF FINANCIAL MEASURES

Financial measures are usually presented as ratios. Ratios are simply one number divided by another to express a relationship. Some examples are bushels per acre, blossoms per flower, and lambs per ewe. Financial measures are also sometimes presented as single dollar values, often called margins, such as working capital (measure 1) and capital replacement and term debt repayment margin (measure 11).

Financial measures can be calculated from past, present, and future figures (Table 8–3). These are often separated into

1. historical information.
2. current information.
3. projected information.

Financial measures can be assessed by comparison to the following:

■ *Other measures.* This allows for a confirmation of conclusions made from evaluating a financial measure. Sometimes the conclusion is changed by comparing it to other measures.

■ *The same measure over time.* This allows for comparison of a business to itself to evaluate the general trend of the business.

■ *The same measures for similar businesses.* Comparisons should generally be made to businesses of a similar size and type. This allows for a more specific evaluation of a business.

■ *The same measures for dissimilar businesses.* Sometimes one wants to compare a business to a dissimilar business. For example, one may want to evaluate whether equity capital would generate a better rate of return if invested in a different type of business.

■ *A known benchmark.* If financial measures differ from industry standards or other benchmarks, a problem area or a strength area exists. Another example of a benchmark is budgeted measures.

## Summary

Financial measures help formulate significant questions, but they do not provide all the answers (Figure 8–7). Financial measures are not a substitute for informed judgment and common sense, which some business owners call "gut instinct." Evaluation through the use of financial measures enhances one's decision-making ability and offers a means to assess the status of an individual business. Selectivity in the choice of measures is imperative, because not every measure is used in every decision-making situation.

### *SUPERVISED EXPERIENCE*

Renewable natural resources projects are popular school-based supervised experiences. A neighboring state park partnered with this school in a renovation effort that required the students to present their plan for approval with state public officials. All aspects of the renovation had to be accounted for.

## TABLE 8-3   TREND ANALYSES*

| Item | For Month and Year Ending | | | | |
|---|---|---|---|---|---|
| **Financial Measures** | | | | | |
| Total assets | | | | | |
| Total liabilities | | | | | |
| Owner equity | | | | | |
| Gross revenue | | | | | |
| Total cash expense | | | | | |
| Net income from operations | | | | | |
| Net business income | | | | | |
| **Liquidity** | | | | | |
| 1. Working capital | | | | | |
| 2. Current ratio | | | | | |
| 3. Cash flow coverage ratio | | | | | |
| **Solvency** | | | | | |
| 4. Debt/asset ratio | | | | | |
| 5. Equity/asset ratio | | | | | |
| 6. Debt/equity ratio | | | | | |
| **Profitability** | | | | | |
| 7. Return on assets | | | | | |
| 8. Return on equity | | | | | |
| 9. Operating profit margin ratio | | | | | |
| **Repayment Capacity** | | | | | |
| 10. Term debt and capital lease coverage ratio | | | | | |
| 11. Capital replacement and term debt repayment margin | | | | | |
| **Financial Efficiency** | | | | | |
| 12. Asset turnover ratio | | | | | |
| 13. Operating expenses ratio | | | | | |
| 14. Depreciation expense ratio | | | | | |
| 15. Interest expense ratio | | | | | |
| 16. Net income from operations ratio | | | | | |

*Source: The National Council for Agricultural Education, Decisions & Dollars, 1995

**FIGURE 8-7**

Establishing a welding business requires streamlined planning and accurate record keeping.

Students formed committees and met regularly with state park officials until the project was completed (Figure 8–8). The project was granted a specific amount of money, and the students had to budget each phase of the renovation so that they would not go over budget. Although the project was a lot of hard work, the students were able to apply what they learned in the classroom and improve their community at the same time.

## CAREER OVERVIEW

Financial management understanding, agricultural experience, and scientific knowledge are all prerequisites for the following careers:

High school diploma: Landscape Assistant; Feed Store Employee

Associate degree: Technical Service Representative; Turf Producer

Bachelor's degree: Food Broker; Export Sales Manager

Advanced degree: Veterinarian; Research and Development Manager

## DISCUSSION QUESTIONS

1. Identify and explain the five criterion areas on that form a complete financial analysis of a business.
2. Categorize the 16 financial measures with the five criterion areas: liquidity, solvency, profitability, repayment capacity, and financial efficiency.

**FIGURE 8-8**

This student and teacher relax after completing their school-based natural resources renovation project.

3. Describe each of the 16 financial measures.
4. Explain the guidelines for applying the use of financial measures.

## SUGGESTED ACTIVITIES

1. Calculate the 16 financial measures for a school-related or personal business.
2. Invite a financial representative to your class and identify which financial measures are utilized most often by his or her firm and why.

## GLOSSARY

**Debt**   Liabilities against a person or business.

**Financial Efficiency**   A measurement of the degree of efficiency in using labor, management, and capital.

**Liquidity**   Consisting of or capable of ready conversion to cash. Measures the ability of a business to meet financial obligations as they come due without disrupting normal ongoing operations.

**Profitability**   A measurement of the extent to which a business generates a profit or net income from the use of labor, management, and capital. It is also the ability to retain earnings and owner equity growth. A business can survive, compete, and progress if its profitability is at a successful level. The profitability and financial efficiency of the business are criteria usually found in an income statement.

**Repayment Capacity**   A measurement of the ability to cover term debt and capital lease payments.

**Revenue**   Income to the business from the sale of goods or services or changes in inventory. It consists of the proceeds received or value created from current business operations.

**Solvency**   Describes a business in sound financial condition that is able to pay all its liabilities.

# 9 Planning and Decision Making

## OBJECTIVES

*After reading this chapter, you will be able to:*

- Describe the decision-making process.
- Identify four purposes of a budget.
- Define variable costs and fixed costs.
- Distinguish variable costs from fixed costs.
- Identify the three main formats for income statement budgets.
- Apply the decision-making process to your business and personal life.
- Identify possible resources available to your business/personal life.
- Identify the primary use of the cash flow budget planning tool.
- List the elements of a cash flow budget.

# BUSINESS PROFILE

**B**ill began a Supervised Agricultural Experience (SAE) program when he first enrolled in agricultural education in the ninth grade. He began work in the shop of River View Lawn Care, a business that sells mowers, trimmers, and other lawn care equipment. In his agricultural education class and in his job, Bill learned to repair small engines. He became so good at repairing, overhauling, and maintaining engines that he opened a shop for himself. The shop made enough money to pay his school expenses and gave him plenty of money to enjoy life.

Bill recently graduated from a two-year postsecondary school where he studied engine repair techniques. Now that he has graduated he has a big decision to make. Although his business has been profitable, he knows that he will have to expand it to be able to support himself and later on a family. Bill has no doubt that if he expands, he can get enough new customers to make the business return a decent profit. His real dilemma is that he has been offered a good job. Clovis's garage is a very large repair shop across town with a branch shop that repairs small engines. The owner, Mr. Clements, has offered Bill a good salary to manage the small engine repair shop. Bill has always wanted to own a business of his own, but he is not sure he can make as much profit with his own shop as he could make with the salary Mr. Clements is offering him.

While Bill was enrolled at the postsecondary school, he took several courses in running a business. One of the courses taught him how to do financial planning and decision making. He began to analyze his problem using a systematic method. First he outlined his goals. He wanted to make at least as much expendable income in his proposed business as he could working for Mr. Clements; so he needed to compare the two alternatives. He knew that, if he took the job, there would be expenses such as costs incurred driving across town each day, the cost of uniforms that Mr. Clements would require, and the deductions such as income tax, social security, and insurance.

Analyzing the costs of owning his own business was more complicated. He needed to know how much more money he would make by taking on new customers and added business. From his classes he knew that increasing the number of customers also increases the cost of doing business. To get a clear view of the expenses and revenues of operating an expanded business, he completed and analyzed the base financial statements such as the balance sheet, income statements, and cash flow statements. He then completed budgets and determined that his expendable income ($2,033) for his own business could be similar to his salary offer ($1,998). Based on this comparison, he decided to stay with his own expanded business.

### Monthly Budget for Bill's Proposed Expanded Business

| Description | Expenses | Revenues |
|---|---|---|
| Building rent | $ 455.00 | |
| Heating and cooling | 220.00 | |
| Part-time help | 360.00 | |
| Interest on loans | 78.00 | |
| Janitorial supplies | 23.00 | |
| Inventory maintenance | 800.00 | |
| Advertising | 56.00 | |
| Tools | 125.00 | |
| **Total Expected Expenses (A)** | **$2,117.00** | |
| Income from repair and maintenance work | | $ 3,300.00 |
| Income from parts sales | | 850.00 |
| **Total Expected Revenue (B)** | | **$4,150.00** |
| **Total Expected Expendable Income (B − A = C)(C)** | | **$2,033.00** |

### Monthly Budget for Bill's Salary Offer

| Description | Expenses | Revenues |
|---|---|---|
| Uniform rental | $ 35.00 | |
| Transportation costs | 125.00 | |
| Tools | 125.00 | |
| **Total Expected Expenses (A)** | **$285.00** | |
| **Monthly Salary (B)** | | **$2,283.00** |
| **Total Income from Salary (B − A = C)(C)** | | **$1,998.00** |

## Introduction

The purpose of this chapter is to learn how a systematic analytical process can be used to make financial decisions. The decision-making process is applicable to every phase of business and personal growth. Establishing realistic **goals** with accurate information is essential when making business decisions and planning for the future. Utilizing all available information prior to becoming obligated to a certain course of action is a goal of this chapter.

## ■ THE DECISION-MAKING PROCESS

The financial decision-making process involves three steps:

1. *planning*, which focuses on preparing a **budget**
2. *deciding*, which involves selecting a course of action to follow
3. *evaluating*, which emphasizes comparing your budget to actual results

The relationships among these steps are shown in Figure 9–1.

## ■ PURPOSES OF A BUDGET

The major financial tool used in decision making is a budget. A budget is a financial plan for acquiring and using your **resources** over a future period.

The purposes of a budget are to:

1. formalize your plans.
2. communicate your plans.
3. evaluate your performance.
4. coordinate your activities.

Costs (**expenses**) are typically categorized by **cost behavior** for planning (budgeting) purposes.

*T*roy asked, "Like the growth of plants, do some costs grow slowly and others quickly?"

*M*s. Green responded, "Troy, that is a good analogy and it is also correct; some costs do increase faster than others."

## ■ VARIABLE AND FIXED COSTS

As the production level of a plant nursery increases, some costs increase and some costs remain constant (Figure 9–2). For example, as you produce more plants, the required amount of soil for containers increases, but the amount of equipment needed (wheelbarrows, hoses, tools, etc.) remains fairly constant.

**FIGURE 9-1**

The decision-making process.

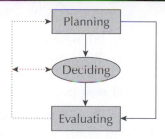

**FIGURE 9-2**

How do these plants differ? Like costs, some plants grow quickly and some grow slowly.

For planning purposes, Troy and Megan must be able to estimate which costs change as production changes (**variable costs**) and which costs remain constant as production changes (**fixed costs**). Applying the concept of variable costs can become difficult, as revealed by the following example.

Assume that the cost of soil per container is $.25 and that Troy and Megan want to look at changes in this variable cost at different production levels. Table 9–1 presents this issue:

**TABLE 9-1    FIXED AND VARIABLE COSTS**

| Number of plants produced | Soil cost per plant | Total soil costs |
|:---:|:---:|:---:|
| 1 | $.25 | $   .25 |
| 20 | $.25 | $ 5.00 |
| 100 | $.25 | $25.00 |

Total soil costs are variable. One difficulty often encountered when applying these concepts is that soil cost per plant is fixed at $.25 per plant. Therefore, a variable cost is described as a variable in total cost. However, it is typically referred to as simply a variable cost.

In addition, soil cost per plant may not be fixed over wide ranges of production. For example, after Troy and Megan started producing more than 100 plants, they could order soil components direct from a wholesaler. As a result, the soil cost per plant decreases because of quantity discounts (Table 9–2).

To deal with this issue, it is helpful to state the **relevant range** of a cost. Relevant range is the range of production in which the behavior of variable and fixed costs remains the same. Table 9–3 presents the four relevant ranges based on figures from Table 9–2.

Applying the concept of fixed costs can become difficult. One reason for this difficulty is demonstrated by the following example. Assume that the cost of equipment (wheelbarrow, shovels, hoes, etc.) is $50 and that Troy and Megan wanted to look at the cost per unit of this fixed cost at different production levels (Table 9–4).

Total equipment costs are fixed. One difficulty often encountered when categorizing costs is that total equipment cost per plant is variable ranging from $50.00 to $.50 per plant in Table 9–4. Therefore, a fixed cost is described as a fixed in total cost. However, it is typically referred to as just a fixed cost.

**TABLE 9-2   DISCOUNTED VARIABLE COSTS**

| Number of plants produced | Soil cost per plant | Total soil costs |
|:---:|:---:|:---:|
| 1 | $.25 | $   .25 |
| 20 | $.25 | $ 5.00 |
| 100 | $.25 | $25.00 |
| 101 | $.20 | $20.20 |
| 301 | $.18 | $54.18 |
| 501 | $.15 | $75.15 |

**TABLE 9-3   RELEVANT RANGES FOR VARIABLE COSTS**

| Relevant range (Number of plants produced) | Soil cost per plant |
|:---:|:---:|
| 1 to 100 | $.25 |
| 101 to 300 | $.20 |
| 301 to 500 | $.18 |
| 501 and greater | $.15 |

**TABLE 9–4    FIXED COSTS AND VARIOUS PRODUCTION LEVELS**

| Number of plants produced | Equipment cost per plant | Total equipment costs |
|:---:|:---:|:---:|
| 1 | $50.00 | $50.00 |
| 20 | $ 2.50 | $50.00 |
| 100 | $  .50 | $50.00 |

**TABLE 9–5    INCREASES IN FIXED COSTS**

| Number of plants produced | Equipment cost per plant | Total equipment costs |
|:---:|:---:|:---:|
| 1 | $50.00 | $ 50.00 |
| 20 | $ 2.50 | $ 50.00 |
| 100 | $  .50 | $ 50.00 |
| 200 | $  .25 | $ 50.00 |
| 301 | $  .33 | $100.00 |
| 501 | $  .20 | $100.00 |

**TABLE 9–6    RELEVANT RANGES IN FIXED COSTS**

| Relevant range (Number of plants produced) | Total equipment costs |
|:---:|:---:|
| 1 to 300 | $ 50.00 |
| 301 and greater | $100.00 |

In addition, total equipment costs may not be fixed over wide ranges of production. For example, if Troy and Megan start producing more than 300 plants, they will probably need more equipment (e.g., soil mix storage containers). Therefore, the total equipment costs will increase (Table 9–5).

To deal with this issue, it is helpful, again, to state the relevant range of a cost. Again, relevant range is the range of production within which fixed and variable costs' behavior remains the same. Table 9–6 presents the four relevant ranges based on Table 9–5.

## ■ BUDGETS

Two types of commonly prepared budgets are the **income statement budget** and the **cash flow budget.** The three main presentation formats of income statement budgets are identified in Table 9–7.

The selected format is primarily dependent on the user's need. The **enterprise budget** format is utilized in the following example. Troy and

## TABLE 9-7   INCOME STATEMENT FORMATS

1. **Partial budget:** Only income statement items that change are presented.

2. **Enterprise budget:** A full income statement is presented for *one segment* or part of a business (e.g., production of a specific item).

3. **Whole business budget:** Full income statements are presented for *all segments* of a business.

Megan want to make lots of money. They are considering ways to generate more cash income for their school-based horticulture business, Sprout About. One alternative that they are considering is to increase their production of Doc's Delite from 100 plants (grown from seed to 1-gallon container size) per year to 200 plants per year. In the past each student in the class sold 100 plants per year at the swap meet, but now they think they can double their **revenue.** Megan has gathered the information listed in Table 9–8 from past records.

Megan and Troy discussed their business idea with the local extension agent, who informed them that Doc's Delite is in high demand because of its low water use and fast growth.

*M*egan said, "If our revenue doubles from 100 to 200 plants, then our profits double from $199 to $398."

*M*s. Green suggested, "Maybe you should prepare a budget."

## TABLE 9-8   TROY'S AND MEGAN'S INCOME STATEMENT

| | Items | Value/unit | Units | Total value |
|---|---|---|---|---|
| *Revenue:* | | | | |
| | Doc's Delite | $4.00 | 100 | 400.00 |
| | | | Total revenue (A) | $400.00 |
| *Variable costs:* | | | | |
| | Soil components | $ .25 | 100 | 25.00 |
| | Containers | $ .75 | 100 | 75.00 |
| | Seeds | $ .01 | 100 | 1.00 |
| | Water | $ .50 | 100 | 50.00 |
| | | | Total variable costs (B) | $151.00 |
| *Fixed costs:* | | | | |
| | Wheelbarrow | | | 50.00 |
| | | | Total fixed costs (C) | $ 50.00 |
| *Net income:* | | | (A − B − C = D) | $199.00 |

Megan and Troy reluctantly prepared the budget shown in Table 9–9 (see Figure 9–3).

Their estimate of **net income** when they simply doubled their $199 figure to $398 was $50 lower than the amount shown on the enterprise budget format, which is $448. The income statement did not indicate that fixed costs are fixed in total in the relevant range of production that they are considering. Troy and Megan need a more systematic approach to help them apply decision making to their business venture (Figure 9–4 and Table 9–10).

**FIGURE 9–3**

Preparing a budget is an appropriate business procedure.

**TABLE 9–9   TROY'S AND MEGAN'S ENTERPRISE BUDGET FORMAT: INCOME STATEMENT**

| | Items | Value/unit | Units | Total value |
|---|---|---|---|---|
| *Revenue:* | | | | |
| | Doc's Delite | $4.00 | 200 | 800.00 |
| | | Total revenue (A) | | $800.00 |
| *Variable costs:* | | | | |
| | Soil components | $ .25 | 200 | 50.00 |
| | Containers | $ .75 | 200 | 150.00 |
| | Seeds | $ .01 | 200 | 2.00 |
| | Water | $ .50 | 200 | 100.00 |
| | | Total variable costs (B) | | $302.00 |
| *Fixed costs:* | | | | |
| | Wheelbarrow | | | 50.00 |
| | | Total fixed costs (C) | | $ 50.00 |
| *Net income:* | | (A − B − C = D) | | $448.00 |

**FIGURE 9-4**

This instructor demonstrates a systematic approach for applying decision-making processes.

**TABLE 9-10     SYSTEMATIC DECISION-MAKING PROCESS***

| Planning | |
| --- | --- |
| | Define the problem. |
| | Gather information about alternative solutions. |
| | Evaluate alternative solutions. |
| Deciding | |
| | Choose alternatives consistent with goals and objectives. |
| | Implement solution. |
| Evaluating | |
| | Analyze the consequences. |
| | Accept the consequences. |

*Source: The National Council for Agricultural Education, DECISIONS & DOLLARS, 1995

## ■ SYSTEMATIC PLANNING

Ms. Green prepared a worksheet (Table 9–11) that helped Megan and Troy perform the **systematic decision-making process** for Sprout About (Figures 9–5 and 9–6).

**TABLE 9–11   TROY'S SYSTEMATIC DECISION-MAKING PROCESS**

1. Define the problem.
   Example: Troy wants to make more money so he can buy a car.

2. Gather information about alternative solutions.
   Example: Take a part-time job at the service station.
   Sell more plants.

3. Evaluate alternative solutions.
   Example: Prepare a budget.

4. Make decisions consistent with goals and objectives.
   Example: Troy's goal is to buy a car. Troy can buy a car sooner if he works at the service station.

5. Implement solution.
   Example: Troy applies for a job at the service station, is offered the job, and accepts the job.

6. Evaluate the consequences.
   Example: Troy is making more money, but he does not find employment at the service station as satisfying as growing plants.

7. Accept the consequences.
   Example: Troy accepts the fact that he has traded a reduced enjoyment of his work for making more money.

**FIGURE 9–5**

Accepting consequences is a key component in the decision-making process. These boys wanted pets, and the consequence of owning a pet is caring for it daily.

**FIGURE 9-6**

Decision-making skills are used continuously during this grooming session.

Troy's and Megan's organized plans assisted them in solving their problems by making timely decisions on how to reach their goals. Yet, before they can reach a goal, they must define what a goal is (Table 9–12). "Goal" is an important word for decision makers. Ms. Green emphasized that guidelines are necessary for setting financial goals (Table 9–13).

Ms. Green shared with Tory and Megan a worksheet for goal setting (Table 9–14), an effective tool in the decision-making process.

The list of techniques in Table 9–15, to be used in reaching one's goals, assisted Troy and Megan in their venture (Figures 9–7 and 9–8).

Ms. Green contemplated selling tens of thousands of Doc's Delight plants. Is the idea consistent with Megan's and Troy's goals? Troy and Megan must

| TABLE 9-12   GOAL* |
| --- |
| Goal: <br> a definite statement that identifies <br> activities, <br> enterprises, <br> and other achievable levels, <br> that you aspire toward. |

*Source: The National Council for Agricultural Education, DECISIONS & DOLLARS, 1995

**TABLE 9–13   FINANCIAL GOAL GUIDELINES**

1. Determine financial goals based on wants and needs.
2. Set a time line and budget for goal accomplishment.
3. Prioritize goals.
4. Keep goals and time line for accomplishment realistic and attainable.
5. Incorporate compatibility.
   a. Business and family activities.
   b. Needs and desires.
6. Make the goal definite and measurable.

**TABLE 9–14   THE GOAL-SETTING PROCESS***

1. Complete the following by listing goals related to your financial business/personal program or project.

   a. Short-term goal: _____

   _____

   b. Intermediate goal: _____

   _____

   c. Long-term goal: _____

   _____

2. List two steps you might take to reach each goal.

   a. 1. _____
      2. _____
   b. 1. _____
      2. _____
   c. 1. _____
      2. _____

3. List two adjustments you may have to make in your business or personal habits to achieve your goals.

   a. _____
   b. _____

*Source: The National Council for Agricultural Education, DECISIONS & DOLLARS, 1995

consider a major question: Do they have the resources to sell tens of thousands of plants? Ms. Green prepared a list of possible resources (Table 9–16) that will help Troy and Megan make decisions. They used a resource **inventory** to accomplish their task of identifying their assets.

**TABLE 9-15   REACHING GOALS**

1. Review goals on a timely basis.
2. Modify goals as resources change or when accomplished.
3. Use time wisely.
4. Organize a work list.

**FIGURE 9-7**

One has to do more than dream about goals. These students aspire to be agricultural scientists, and they are doing more than dreaming; they are enrolled in a biotechnology class.

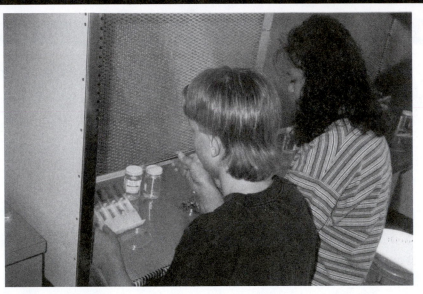

**FIGURE 9-8**

These young men have a goal to become veterinarians.

**TABLE 9–16    RESOURCE INVENTORY\***

A. Land

B. Buildings

C. Labor

    i.  Types of jobs

    ii.  Career development

    iii.  Education required

D. Machinery

E. Capital

F. Management

G. Facilities and fixtures

H. Technology

\*Source: The National Council for Agricultural Education, Decisions & Dollars, 1995

As Megan completed the resource list, Troy observed that many of the resources are assets presented on the balance sheet.

# ■ CASH FLOW BUDGET

What if Troy and Megan do not have enough cash or **credit** to sell tens of thousands of plants? The need for **cash flow** planning arises. The primary use of a cash flow budget is to estimate when cash will need to be borrowed and when it can be repaid.

More specifically, a cash flow budget allows the students to:

1. develop a borrowing and debt repayment plan.
2. rearrange purchases and scheduled debt repayments to minimize borrowing.
3. combine business and personal financial affairs into one complete plan.
4. spot an imbalance between current and non-current credit.

The parts of a cash flow budget in the following list are presented on the student's cash flow budget (Table 9–17).

1. cash income
2. cash expenses
3. borrowed funds needed
4. loan repayments
5. ending cash balance (total cash available)

Because Troy and Megan do not have any accruals, their cash flow is the same as the accrual basis presented on their income statement. However, they must purchase their supplies and equipment before they sell any plants.

**TABLE 9–17    TROY'S AND MEGAN'S CASH FLOW BUDGET***

|  | Time Period 1 | Time Period 2 |
|---|---|---|
| Beginning Cash Balance | $ 0.00 | $ 0.00 |
| Cash Inflow:<br>        Doc's Delite<br>Total Cash Inflow | $ | $ 400.00<br>400.00 |
| Cash Outflow:<br>        Soil components<br>        Containers<br>        Seeds<br>        Water<br>        Wheelbarrow<br>Total Cash Outflow | $ 25.00<br>75.00<br>1.00<br>50.00<br>50.00<br>201.00 | $<br><br><br><br><br>0.00 |
| Cash Balance (total inflow – total outflow) | $ (201.00) | $ 400.00 |
| Borrowed funds needed | $ 201.00 | $ 0.00 |
| Loan repayments (principal and interest) | $ | $ 201.00 |
| Ending Cash Balance<br>(cash balance + borrowed funds – repayments) | $ 0.00 | $ 199.00 |
| Debt Outstanding | $ 201.00 | $ 0.00 |

*Source: The National Council for Agricultural Education, DECISIONS & DOLLARS, 1995

# Summary

Troy and Megan recognized that a good understanding of fixed and variable costs is important when they began the planning and decision-making process. Furthermore, applying the financial decision-making skills to both business and personal situations can enhance the outcomes of their decisions. A complete and accurate assessment of the business's resources must be available for the best possible decisions. Finally, Megan and Troy concluded that utilizing a cash flow budget in the decision-making process is a powerful tool.

## SUPERVISED EXPERIENCE

Many young people have a great love for animals, but most people do not know how to properly care for them. These students have a school-based animal science laboratory supervised experience that emphasizes animal care (Figure 9–9). A wide variety of animals are kept at the school, but the big favorites are the horses. Students must earn the right to adopt an animal by obtaining high grades, attending classes, and earning outstanding citizenship scores. Riding and caring for the large animals is available only to students who have successfully completed the beginning agriscience courses and who have kept up-to-date and accurate record books.

**FIGURE 9–9**

Students learn proper hoof care as part of their school-based animal laboratory project.

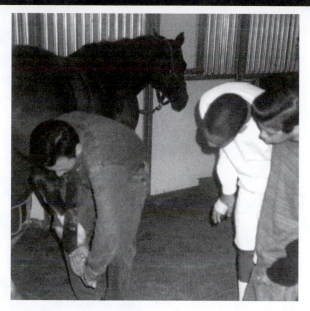

The high standards ensure that responsible students are eligible to take care of the large animals (cattle, horses, sheep, or pigs) (Figure 9–10). Students must continue to keep and maintain financial and production records on their adopted animals throughout their high school career. The students pay a fee each year for the use of the facilities. When students graduate, they are expected to return their adopted animal to the school or replace its with one of its offspring. Most students continue their education at the local community college or at the land grant university.

## CAREER OVERVIEW

Students with nonagricultural living arrangements can gain agricultural experience through high school agriscience programs (Figure 9–11). Career preparation is a high priority in agriscience programs because agriculture, business, and science are emphasized in the curriculum. The following careers are available to agriscience students:

High school diploma: Vegetable Market Seller; Dairy Milker

Associate degree: Dairy Processing Technician; Teaching Aide

Bachelor's degree: Aquaculturalist; Agribusiness Manager

Advanced degree: Market Analyst; Entomologist

## DISCUSSION QUESTIONS

1. Explain the three-step decision-making process.
2. Identify and describe the four purposes of a budget.

**FIGURE 9-10**

These girls are making many decisions as they care for this horse's leg.

**FIGURE 9-11**

Sharing in the joy of discovery is a goal of Ag(riculture) in the Classroom and a great way to encourage young children to enjoy learning and making decisions. [Ag(riculture) in the Classroom is sponsored by USDA-CSREES and managed through a cooperative agreement with Utah State University.]

3. Distinguish between variable and fixed costs.

4. Contrast between Megan's and Troy's total soil costs and total equipment costs.

5. Differentiate between Megan's and Troy's possible net income as generated by the income statement ($398) and by the enterprise budget format ($448).

6. Appraise the rationale for identifying adjustments in the goal-setting process.

7. Compare the purpose of a resource inventory and a balance sheet.

8. Identify and explain the benefits of a cash flow budget.

## SUGGESTED ACTIVITIES

1. Identify a problem and complete the systematic decision-making process.

2. Employ the goal-setting process for a personal, school, business, and/or family activity.

3. Complete a cash flow budget for a personal, school, business, and/or family activity.

4. Compare methods of capital resource acquisition.

## GLOSSARY

**Budget**   Formal written or unwritten plan that projects the use of assets for a future time.

**Cash Flow**   Receipts and disbursements in the business; determines the cash flow budget.

**Cash Flow Budget**   Statement of projected expenses and receipts associated with a particular farm or business plan.

**Cost Behavior**   How a cost responds to changes in production volume.

**Credit**   Means of obtaining goods or services now by promising to pay at a later date.

**Enterprise Budget**   Full income statement presented for one segment or part of a business (e.g., production of a specific item).

**Expense**   Cost of goods or services involved in producing a product or service.

**Fixed Costs**   Costs that remain mostly constant as production changes within a relevant range.

**Goals**   Definitive statement that identifies activities, enterprises, or other achievable levels that you aspire toward.

**Income Statement Budgets**   Budgets that use the income statement format when created. Examples include the partial, enterprise, and whole business budgets.

**Inventory**   Listing and valuation of all business assets.

**Net Income**   Revenue minus expenses. Calculated by matching revenues with the expenses incurred to create those revenues, plus the gain or loss on the normal sales of capital assets. *See also* profit.

**Partial Budget**   A financial statement containing only income statement items that change.

**Relevant Range**   Range of production in which variable and/or fixed costs' behavior remains the same.

**Resources**   Sources of supply or support.

**Revenue**   Income to the business from the sale of goods or services or changes in inventory. It consists of the proceeds received or value created from current business operations.

**Systematic Decision Making Process**   Steps include: (1) define the problem, (2) collect information, (3) evaluate alternative solutions, (4) make decisions consistent with the goals and objectives, (5) implement a solution, (6) evaluate consequences, and (7) accept the consequences.

**Variable Costs**   Costs that change as production changes within a relevant range.

**Whole Business Budget**   Full income statements are presented for all segments within the business (e.g., production of a specific item).

# 10

# Business Borrowing and Investing

## OBJECTIVES

*After reading this chapter, you will be able to:*

- Distinguish the two main types of capital.
- Integrate borrowing with the decision-making process.
- Describe five main tools for evaluating capital needs.
- Apply tools for evaluating capital needs to your business and personal life.
- Describe the types of loans available to personal and business borrowers.
- Compare two methods for calculating the cost of borrowing.
- Apply the compound interest method to inflation.

# BUSINESS PROFILE

Remember from the previous chapter that Bill has decided to expand his small engine repair business so that he will have enough money from the operation for living expenses. His next big decision is how fast and how much he wants to expand. Because there is a high demand for his services, he can considerably expand his business.

However, Bill also realizes that if he tries to expand too much, too fast, he may encounter unacceptable financial risks in the business. He knows that to expand he will have to borrow some money. He is not sure just how much he can or should borrow from the bank to expand the business. In a financing class he took, Bill learned that he would have to combine his current capital with borrowed capital for the operation to grow. The loan officers at the bank will help Bill calculate a debt-to-equity ratio that will assist him in determining the range of funds he can borrow. This step will tell how much capital or equity Bill will have in the business compared to the debt or how much money the bank invests in the business.

After deciding that he will borrow $15,000 over a period of three years, Bill wisely decides to shop around for the best rate of interest on a loan. He finds that one bank is offering to loan him the money at 12% and another bank will loan the money for 10%. Further investigation reveals that the bank offering the 10% rate is charging compound interest and the bank with the 12% interest is charging simple interest. Which of the two rates would you advise Bill to take?

**SOLUTION:** The total repayment of the $15,000.00 loan for three years at 12% simple interest amounts to $20,400. At 10% compound interest, the total repayment amounts to $23,784. Bill should borrow the money at 10% compound interest rate rather than 12% simple interest rate.

| Loan Principal | Length of Loan | Interest Total | Loan Payoff |
|---|---|---|---|
| $15,000.00 | 3 years | 12% simple | $20,400.00 |
| $15,000.00 | 3 years | 10% compound (1.331) | $19,965.00 |
| Difference in the two interest rates: $435.00 | | | |
| Simple interest or add-on interest method: ($15,000.00)(3 years)(12%) + ($15,000.00) = $20,400.00  Principal × length of loan × annual interest rate + original principal = total loan payoff. | | | |
| Compound interest method: ($15,000.00)(1.331) = $19,965.00  Principal × compound interest value (see Table 9—4) = total loan payoff. | | | |

## Introduction

The purpose of this chapter is to make practical and efficient use of borrowed **capital.** Because of the vast amounts of capital needed for today's agribusinesses, it is very important that students learn that capital can be another input in the business process and that it should be treated as such (Figure 10–1).

Troy and Megan learned that the cash flow budget is a valuable tool for planning and decision making. They copied their budget for the first two time periods onto a six-period schedule (Table 10–1). They added a line for **interest** as a cash outflow because they realized that business borrowing carries an additional cost called "interest expense."

## ■ CAPITAL

Capital includes cash or liquid savings, machinery, livestock, buildings, and other assets that have a useful life of more than one year and that meet a specific minimum value. These are all assets presented on the balance sheet. Capital can also be separated into two categories, existing and borrowed. Existing capital is presented as equity on the balance sheet. Borrowed capital is presented as a liability on the balance sheet (Figure 10–2).

**FIGURE 10–1**

Obtaining capital, even for a small nursery project, can be difficult without accurate records and a management plan.

**TABLE 10–1   TROY'S AND MEGAN'S CASH FLOW STATEMENT***

| Item | Period 1 | Period 2 | Period 3 | Period 4 | Period 5 | Period 6 |
|---|---|---|---|---|---|---|
| Beginning Cash Balance | 0.00 | 0.00 | | | | |
| Cash Inflow: | 0.00 | 400.00 | | | | |
| Cash Outflows:<br>　Soil components<br>　Containers<br>　Seeds<br>　Water<br>　Wheelbarrow<br>　Interest | <br>25.00<br>75.00<br>1.00<br>50.00<br>50.00 | | | | | |
| Cash Balance | (201.00) | 400.00 | | | | |
| Borrowed Funds Needed | 201.00 | 0.00 | | | | |
| Loan Repayments | | 201.00 | | | | |
| Ending Cash Balance | 0.00 | 19.00 | | | | |
| Debt Outstanding | 201.00 | 0.00 | | | | |

*Source: The National Council for Agricultural Education, DECISIONS & DOLLARS, 1995

**FIGURE 10–2**

Capital can originate from existing sources such as savings, or it can be borrowed as in a loan.

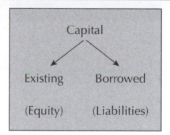

Megan wondered, "When our cash flow budget shows that we need borrowed funds, do we create a liability on our balance sheet?"

Ms. Green was impressed at how Megan's comment demonstrated her understanding of the connection between the balance sheet and the cash flow budget. This understanding is an important resource for the management of a successful business. She answered, "Yes, you are absolutely correct."

## ■ BORROWING AND DECISION MAKING

As demonstrated in Figure 10–3, the financial decision-making process involves three primary stages: planning, deciding, and evaluating.

> *T*roy commented, "I remember, we prepared a cash flow budget during the planning stage. Then I used the budget to evaluate my plans, and finally I took a job at the service station during the decision stage."

Ms. Green drew the thinner lines on the diagram in Figure 10–3 to show Troy how his comment followed the decision-making process.

Troy does not enjoy his job at the service station; he would rather find some capital so that he and Megan could enlarge Sprout About. Ms. Green pointed out to Troy that he is refining his financial goals using the guidelines presented in Table 10–2. Specifically, he is incorporating the compatibility of his desire to enjoy his work and his need to make lots of money so that he can attain his goal of buying a car.

> "*I* understand, I continued the decision-making process by evaluating my decision to work in a service station and decided to perform more planning for my capital needs," Troy stated.

**FIGURE 10–3**

The thin line indicates the decision-making process that Troy used when he decided to work at the service station.

**TABLE 10-2    FINANCIAL GOAL GUIDELINES\***

1. Determine financial goals based on wants and needs.

2. Set a time line and budget for goal accomplishment.

3. Prioritize goals.

4. Keep goals and time line for accomplishment realistic and attainable.

5. Incorporate compatibility.

   a. Business and family activities.

   b. Needs and desires.

6. Make the goal definite and measurable.

*Source: The National Council for Agricultural Education, DECISIONS & DOLLARS, 1995

**FIGURE 10-4**

The dashed line illustrates that Troy's decision-making process utilizes all three components: planning, deciding, and evaluating.

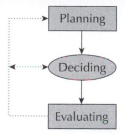

Ms. Green drew the dashed lines on Figure 10–4 to show Troy how his comment followed the decision-making process.

Ms. Green concluded by stating that, with respect to capital needs, two basic questions must be answered.

1. How much total capital to use?
2. How should limited capital be allocated among its many potential uses?

## ■ TOOLS FOR EVALUATING CAPITAL NEEDS

The cash flow budget is one of the primary tools for evaluating how much capital is needed. This budget presents the calculation for the minimum amount of borrowed capital (funds) needed, as demonstrated in Table 10–3. Additional capital could be necessary to provide a cushion in case the budgeted cash flow does not occur as soon as planned.

**TABLE 10-3    TROY'S AND MEGAN'S CASH FLOW BUDGET***

|  | Period 1 | Period 2 |
|---|---|---|
| Beginning Cash Balance | $            0.00 | $            0.00 |
| Cash Inflow<br>    Doc's Delite<br>Total Cash Inflows | $ | $          400.00<br><br>          400.00 |
| Cash Outflow:<br>    Soil components<br>    Containers<br>    Seeds<br>    Water<br>    Wheelbarrow<br>Total Cash Outflow | $          25.00<br>          75.00<br>           1.00<br>          50.00<br>          50.00<br>         201.00 | $<br><br><br><br><br><br>           0.00 |
| Cash Balance (total cash inflow – total cash outflow) | $         (201.00) | $          400.00 |
| Borrowed Funds Needed | $         201.00 | $            0.00 |
| Loan Repayments (principal and interest) |  | $        $201.00 |
| Ending Cash Balance<br>(cash balance + borrowed funds – repayments) | $           0.00 | $          199.00 |
| Debt Outstanding | $         201.00 | $            0.00 |

*Source: Based on information and forms from The National Council for Agricultural Education, DECISIONS & DOLLARS, 1995

---

*T*roy thought, "I can borrow more money than the $201 shown on Table 10–3. I'll be able to buy my car and help Megan enlarge Sprout About."

---

Ms. Green considered Troy's comment and recognized that he may be reflecting on economic principles that encouraged leveraging. Some businesses and individuals borrow heavily to **leverage** what they currently have with the hope of gaining higher profits in the future and satisfying more of their needs. There is also a lot more risk associated with this economic philosophy. Ms. Green decided that it was time to present tools to:

1. measure leverage.

2. evaluate risk.

3. maximize profits.

4. satisfy multiple goals.

5. ration capital.

**FIGURE 10-5**

Leverage is measured by the debt-to-equity ratio.

## Leverage and Debt/Equity Ratio

Leverage =

A measure of the degree to which the business is financed by external resources.

measured by:

$$\text{Debt/Equity Ratio} = \frac{\text{total liabilities}}{\text{total equity}}$$

Source: The National Council for Agricultural Education, DECISIONS & DOLLARS, 1995

## MEASURING LEVERAGE

*M*egan commented, "The leverage concept is like needing a large piece of lumber to move a large boulder and a small piece of lumber to move a small boulder."

Liabilities are like lumber used to move landscape rock, and equity is like the rock. In comparison to short lumber, longer lumber provides greater leverage (which allows the rock to be more easily moved). Similarly, the larger the liabilities are in relation to equity, the greater the leverage is (Figure 10–5).

Both liabilities and equity are considered external funding sources. One difference is that liabilities are borrowed capital and equity is usually existing capital. Therefore, leverage is increased with higher borrowed capital in relation to existing capital.

## EVALUATING RISK

Leverage and risk are closely associated. The same tool is used to calculate leverage and to evaluate risk, the debt-to-equity ratio (Figure 10–6). An increase in the debt-to-equity ratio to 1.5 generally is associated with a heightened risk of losing **equity capital,** compared to a debt-to-equity ratio of, say, .5.

## MAXIMIZING PROFITS

To maximize profits, borrowed capital must be made to produce revenue. The decision rule is used to evaluate how borrowing capital will change net income (Figure 10–7).

**FIGURE 10-6**

**Equity capital equals total assets minus total liabilities.**

## Principle of Increasing Risk

If:     debt/equity ratio ↑

Then: risk of losing equity capital ↑

debt ÷ equity ratio = total liabilities ÷ total equity

examples        debt/equity ratio

Example A    $= \dfrac{\$23,000.00}{\$64,000.00} = 0.5$

Example B    $= \dfrac{\$96,000.00}{\$64,000.00} = 1.5$

Source: The National Council for Agricultural Education, DECISIONS & DOLLARS, 1995

**FIGURE 10-7**

**Understanding the relationship between MVP and MIC is crucial when making business decisions.**

## Decision Rule

MVP—Marginal Value Product
  (additional revenue from additional input)

MIC—Marginal Input Costs
  (the cost of additional input)

Break-Even Point:

  where MVP = MIC

when MVP > MIC = *more net income* with more input
when MVP < MIC = *less net income* with more input

Source: The National Council for Agricultural Education, DECISIONS & DOLLARS, 1995

In summary, when the **marginal input cost (MIC),** including the costs of borrowing (interest expense), is less than the **marginal value product (MVP),** or incremental revenue produced from the expenditure of MIC, then net income increases. When MIC, including costs of borrowing (interest expense), is greater than MVP, then net income decreases.

## SATISFYING MULTIPLE GOALS

Troy is trying to satisfy two primary goals:

1. enjoyment of his work
2. accumulation of enough money to buy a car

A common way to work toward both goals is to follow the **equal marginal principle** (Figure 10–8).

> *T*roy questioned, "Should I spread my limited borrowed funds [input] so that I can reach toward both of my primary goals at the same time?"
>
> *"T*hat's a great idea," remarked Ms. Green.
>
> *T*roy became very excited that he may be able to enjoy his work and have enough money to buy a car.

## CAPITAL RATIONING

Available capital can be limited by both external and internal factors. For example, Troy is probably not able to borrow capital from a bank (an external source) because banks usually limit their borrowing to established businesses.

Troy may be able to borrow capital from a family member, but he may decide that to borrow capital is too risky (an internal factor).

**FIGURE 10-8**

The equal marginal principle is a tool used to work toward the successful completion of goals.

### Equal Marginal Principle

(Equal "Changes" Principle)

Limited input allocated among alternatives so that Marginal Value Product (MVP) is equal in all units.

In other words . . . . . . . . . . . . . . . . . . . . . . . . . . . . . . . . . . . . . . . . . . . . . . . . . .

Spread the input around so that MVP is equal in all your enterprises.

Source: The National Council for Agricultural Education, DECISIONS & DOLLARS, 1995

## ■ LOAN OPTIONS

There are various categories, criteria, and factors to consider when borrowing capital (Figure 10–9). Some of the major options follow.

### LOAN LENGTH

Loans that will be paid back in full within one year are short-term loans. Loans that will be paid back over more than one year are long-term loans.

### LOAN USE

Loans are used for three general purposes:

1. *Real estate loans.* Land and/or buildings are used as **collateral.**
2. *Nonreal estate loans.* Loans other than real estate loans are made to businesses.
3. *Personal loans.* Nonbusiness loans may be used to purchase personal assets.

### LOAN SECURITY

Secured loans have assets that serve as collateral (Figure 10–10). Collateral are assets that are pledged or mortgaged to secure repayment of a loan. Unsecured loans (also called signature loans) have no collateral and are based on good credit and loan history.

**FIGURE 10-9**

A loan enabled these students to purchase a large quantity of feed at great savings.

**FIGURE 10-10**

The cattle chute and scale were purchased with borrowed money. Later when they were paid off, they served as collateral on other loans.

## LOAN REPAYMENT SCHEDULE

Loan payments can be separated into two categories.

1. **Principal** is the original loan value, the amount borrowed from the lender (Figure 10–11).
2. Interest is the amount paid to the lender for the use of money.

**FIGURE 10-11**

Learning how to calculate principal and interest payments is an integral part of this student's agriscience education.

**FIGURE 10–12**

Teaching about amortized repayment schedules is simplified with personalized instruction.

Loan repayments can generally be classified into two categories.

1. *Single payment.* All the principal is payable in one lump sum when the loan is due.
2. **Amortized.** The principal and interest payments are made periodically.

Each payment can have equal principal, meaning that

- the same amount of principal is due with each payment, along with interest on the unpaid balance.
- the principal balance decreases, interest decreases, and each total payment decreases.
- the total of the payment for each period is lower than that of the previous payment.

Payments can also have an equal total, meaning that

- the early payments represent less principal and more interest than later payments.
- successive payments contain more principal and less interest.
- the total amount of payment for each period is the same (Figure 10–12).

## ■ SOURCES OF LOAN FUNDS

Commercial banks have traditionally supplied consumer loans on large-dollar purchases such as automobiles. Banks generally prefer customers with established credit histories. Therefore, if you are a first-time borrower, you have to pursue another lending source.

**FIGURE 10-13**

The local agriscience alumni loan service allowed this young man the opportunity to purchase a sheep.

Other lending sources are

- the Farm Credit System.
- life insurance companies.
- the Farmers Home Administration (FHA, a program sponsored by the federal government).
- individuals (such as friends or relatives).
- merchants and dealers (who might lend you money by opening an account with you; for example, Megan and Troy may be able to pay their supplier of soil containers 30 days after they receive the containers).
- a credit union (a cooperative association that accepts savings deposits and makes small loans to its members).
- other sources (Figure 10–13).

## ■ COST OF BORROWING

In the borrowing world, you encounter two basic ways of calculating interest, simple and compound.

1. The **simple interest** method is also called "add-on interest method." Interest is calculated by using the original principal for the entire time period, three years in the example in Figure 10–14.

2. In the **compounding** method, interest is based on the declining principal balance over the length of time the amount is borrowed (Figure 10–15).

**FIGURE 10–14**

> The simple interest method is also called the "add-on interest method" because interest is added onto the amount borrowed.

The formula for calculation is:

Principal × Time (usually in years) × Annual Interest Rate

Assuming $1,000.00 is borrowed for three years at 12% interest, you compute $360.00 of interest as $1,000.00 × 3 years × 12%.

The total repayment at the end of year three of $1,360.00 is computed as $1,000.00 plus $360.00.

**FIGURE 10–15**

> The compounding method of calculating interest results in higher payments.

Assuming $1,000.00 is borrowed for three years at 12% interest, you compute the total repayment at the end of year three of $1,405.00 as $1,000.00 × 1.405 (Compound Interest Table 9–4 value for 3 years × 12%).

Of the two methods, you pay more interest with compounding. The simple interest method assumes that the principal balance remains at, say, $1,000 throughout the life of the loan, whereas the compound interest method accrues interest on interest, resulting in the principal balance's increasing over time. The borrower therefore pays more interest, and the lender earns a higher return on the investment.

## ANNUAL PERCENTAGE RATE

The **annual percentage rate (APR)** under the compounding method is more than 12%. In 1969, the federal government passed the truth-in-lending law, which requires a lending institution to tell the borrower, in writing, what the actual interest rate (APR) is. This interest rate tells the borrower the true cost of the loan.

## INFLATION

**Inflation** is an application of compound interest that affects everyone. The compound interest table can be used to calculate future prices adjusted for inflation (Table 10–4). For example, assuming an inflation rate of 4%, grain that sells for $100 today would sell for $148 ($100 × 1.480) ten years from now.

## Summary

Troy and Megan learned that adding capital to their business, Sprout About, can come from existing sources or from loans. However, the decision to borrow has to be made using the decision-making process and by evaluating

## TABLE 10–4   COMPOUND INTEREST TABLE OR FUTURE VALUE OF A $1 INVESTMENT*

| | Interest Rate | | | | | | |
|---|---|---|---|---|---|---|---|
| Years | 4% | 5% | 6% | 7% | 8% | 10% | 12% |
| 1 | 1.040 | 1.050 | 1.060 | 1.070 | 1.080 | 1.100 | 1.120 |
| 2 | 1.082 | 1.103 | 1.124 | 1.145 | 1.166 | 1.210 | 1.254 |
| 3 | 1.125 | 1.158 | 1.191 | 1.225 | 1.260 | 1.331 | 1.405 |
| 4 | 1.170 | 1.216 | 1.262 | 1.311 | 1.360 | 1.464 | 1.574 |
| 5 | 1.217 | 1.276 | 1.338 | 1.403 | 1.469 | 1.611 | 1.762 |
| 6 | 1.265 | 1.340 | 1.419 | 1.501 | 1.587 | 1.772 | 1.974 |
| 7 | 1.316 | 1.407 | 1.504 | 1.606 | 1.714 | 1.949 | 2.211 |
| 8 | 1.369 | 1.477 | 1.594 | 1.718 | 1.851 | 2.144 | 2.476 |
| 9 | 1.423 | 1.551 | 1.689 | 1.838 | 1.999 | 2.358 | 2.773 |
| 10 | 1.480 | 1.629 | 1.791 | 1.967 | 2.159 | 2.594 | 3.106 |
| 11 | 1.539 | 1.710 | 1.898 | 2.105 | 2.332 | 2.853 | 3.479 |
| 12 | 1.601 | 1.796 | 2.012 | 2.252 | 2.518 | 3.138 | 3.896 |
| 13 | 1.665 | 1.886 | 2.133 | 2.410 | 2.720 | 3.452 | 4.363 |
| 14 | 1.732 | 1.980 | 2.261 | 2.579 | 2.937 | 3.798 | 4.887 |
| 15 | 1.801 | 2.079 | 2.397 | 2.759 | 3.172 | 4.177 | 5.474 |
| 16 | 1.873 | 2.183 | 2.540 | 2.952 | 3.426 | 4.595 | 6.130 |
| 17 | 1.948 | 2.292 | 2.693 | 3.159 | 3.700 | 5.055 | 6.866 |
| 18 | 2.026 | 2.407 | 2.854 | 3.380 | 3.996 | 5.560 | 7.690 |
| 19 | 2.107 | 2.527 | 3.026 | 3.617 | 4.316 | 6.116 | 8.613 |
| 20 | 2.191 | 2.653 | 3.207 | 3.870 | 4.661 | 6.728 | 9.646 |
| 21 | 2.279 | 2.786 | 3.400 | 4.141 | 5.034 | 7.400 | 10.804 |
| 22 | 2.370 | 2.925 | 3.604 | 4.430 | 5.437 | 8.140 | 12.100 |
| 23 | 2.465 | 3.072 | 3.820 | 4.741 | 5.871 | 8.954 | 13.552 |
| 24 | 2.563 | 3.225 | 4.049 | 5.072 | 6.341 | 9.850 | 15.179 |
| 25 | 2.666 | 3.386 | 4.292 | 5.427 | 6.848 | 10.835 | 17.000 |
| 30 | 3.243 | 4.322 | 5.743 | 7.612 | 10.063 | 17.449 | 29.960 |
| 35 | 3.946 | 5.516 | 7.686 | 10.677 | 14.785 | 28.102 | 52.800 |
| 40 | 4.801 | 7.040 | 10.286 | 14.974 | 21.725 | 45.259 | 93.051 |
| 45 | 5.842 | 8.985 | 13.765 | 21.002 | 31.920 | 72.891 | 163.988 |
| 50 | 7.107 | 11.467 | 18.420 | 29.457 | 46.902 | 117.391 | 289.002 |

*Source: The National Council for Agricultural Education, Decisions & Dollars, 1995

capital needs. When this was done, they had to decide on the available loan options and loan sources. Finally, Troy and Megan determined their repayment schedule.

## SUPERVISED EXPERIENCE

When this group of students began their customized greenhouse construction business, they determined that if their business goal was to be realized, then they had to borrow money (Figure 10–16). Their goal was to finance a trip for all of them to the National FFA Convention. The team leaders quickly prepared a budget, decided on individual responsibilities, and developed a time line.

Seven portable greenhouses had to be constructed in three months if the team was going to meet its goal. Teamwork was the key and all members built the greenhouses, but they all had individual jobs. One individual kept track of inventory. Another student formulated a marketing plan and took orders. Still another student was in charge of daily scheduling. The team leaders maintained the records and oversaw the entire operation (Figure 10–17).

The team completed construction on the greenhouses on time, and they spent an entire weekend delivering them to their new owners. Their reward, the trip to the National FFA Convention, is something they will remember for the rest of their lives, as well as the skills they learned in planning, organizing, and constructing the greenhouses.

**FIGURE 10–16**

These students borrowed money to purchase materials and supplies for this school greenhouse.

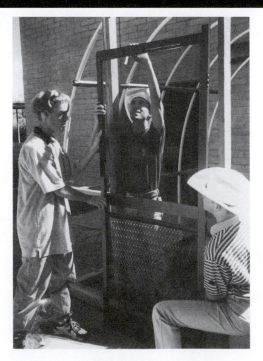

**FIGURE 10–17**

The team leaders had to personally visit with the loan officer prior to loan approval.

## CAREER OVERVIEW

The students in the greenhouse construction business had to have business, agricultural construction, and scientific knowledge to be able to successfully complete their task (Figure 10–18). Careers now available to them include

High school diploma: Nursery Products Grower; Fruit Harvester

Associate degree: Produce Clearing House Specialist; Landscape Contractor

Bachelor's degree: Greenhouse Manager; Horticulture Research Technician

Advanced degree: University Horticulture Scientist; Rangeland Scientist

## DISCUSSION QUESTIONS

1. Distinguish between the two main types of capital.
2. Discuss how borrowing is integrated with the decision-making process.
3. Describe the five main tools for evaluating capital needs.
4. Describe the types of loans available to personal and business borrowers.
5. List and describe eight sources of loan funds.
6. Compare the two methods for calculating the cost of borrowing.
7. Compare the compound interest method to inflation.

**FIGURE 10-18**

Thoroughly preparing a business portfolio is the goal of this student.

## SUGGESTED ACTIVITIES

1. Apply the tools for evaluating capital needs to your business and personal life.

2. Appraise the current financial borrowing status of a school-related entity.

3. Compute and compare the cost of borrowing money for a car utilizing a variety of methods as offered by car dealerships and lending institutions that specialize in car loans.

4. Visit a bank or financial institution loan officer, obtain a loan application, and ask for guidance and advice on completing the form. Prepare a report that outlines the loan officer's suggestions for preparing an application, including identifying the reasons that some applications are approved for borrowing and others are not.

## GLOSSARY

**Amortized**   Principal payments plus periodic interest on loans. Amortized loans are in the form of equal principal payments or equal total payments.

**Annual Percentage Rate (APR)**   Actual interest rate expressed on an annual basis. Also known as effective rate.

**Capital**   Cash or liquid savings, machinery, livestock, buildings, and other assets that have a useful life of more than one year and that meet a specified minimum value.

**Collateral**   Assets pledged or mortgaged to secure the repayment of a loan.

**Compounding**   Interest received from an investment. Interest is added to the principal and interest is paid again on the total sum. It is a method of calculation that can be used to determine the value.

**Equal Marginal Principle**   Input should be allocated among alternative uses in such a way that the marginal value products of the last unit are equal in all uses.

**Equity Capital**   Existing capital investment.

**Inflation**   Prices go up; the value of money goes down.

**Interest**   Amount of funds paid to the lender for the use of money.

**Leverage**   Measure of the extent to which debt capital is being combined with equity capital.

**Marginal Input Cost (MIC)**   The cost of additional input.

**Marginal Value Product (MVP)**   Additional revenue received from additional input.

**Principal**   Amount of money borrowed or invested.

**Simple Interest**   Principal $\times$ Time in years $\times$ Annual rate of interest.

# 11

# Taxes

## OBJECTIVES

*After reading this chapter, you will be able to:*
- Prepare the documents necessary to receive a paycheck.
- Calculate take-home pay.
- List the qualifications for filing a state and federal income tax returns.
- Keep the records required to file an accurate personal income tax return.
- Distinguish between tax-deductible and non-tax–deductible business expenses.
- Distinguish between a deductible expense and a capital purchase.
- Manage income to maximize after-tax income.

## KEY TERMS

| | | |
|---|---|---|
| Audit | I-9 | Social Security |
| Depreciation | IRS | Take-home Pay |
| Exemption | Net Pay | Tax Bracket |
| FICA | Pay Stub | W-2 |
| Gross Pay | Schedule | W-4 |
| Income Tax | Self-employment Tax | |

# BUSINESS PROFILE

Our friend Bill has now begun operating his expanded small engine repair business. The operation is going very well and Bill is making more money than he had anticipated. When he began his enterprise as a Supervised Agricultural Experience (SAE) program several years ago, he began to file income tax returns on the money he made with his part-time job. Even though his income was small compared to what he is now making, he was still required by law to file an income tax return to both the state and the federal governments. At that time, his employer simply deducted the proper amount of state and federal taxes from each of Bill's paychecks to cover the cost of the taxes. Usually, Bill received a refund from the amount of taxes he paid. Because his income was so small, he could file on a very simple form. This form took what is known as a standard deduction that did not take any "cost of operating" deductions into account. The form was simple to fill out, took little time, and gave Bill the means to receive his refund.

Later, when Bill opened his own business, he had to pay his own taxes because he no longer had an employer to withhold taxes from his paycheck. This meant that Bill had to keep meticulous records on the amount of money he received for the services he performed for his customers. Even though he owned his own business, he still did not make enough income to use any forms other than the one that took a standard deduction.

When Bill expanded his business, he started to make enough profit to warrant using a longer filing form. This form allowed him to deduct expenses incurred in operating his business. Bill realized that supplies bought for the business affected the amount of net profit his business made. Such items as cleaning solvents, welding supplies, paint, oils, sandpaper, and other expendable supplies were all allowed by the government as operating expenses. Taking these deductions reduced the amount of money used to figure the amount of taxes he owed.

As the business grew, Bill began to pay more and more taxes. One day he was talking with his aunt, who was an accountant, about the high rate of taxes he had to pay. His asked his aunt if he was taking all of the exemptions or deductions to which he was entitled. Bill began to tell her about all the things he entered as deductions on his tax forms. He aunt asked if he had thought about such things as depreciation on his equipment, the expenses incurred with his delivery truck, the rent on the building he used, or the utilities for the business. Bill replied that he never knew these items could be used as deductions. His aunt assured him that it might be possible that all of these items and more could be legal deductions from the business's profit.

Bill realized that he needed the help of an expert in computing taxes. That year he used a professional tax consultant to figure his taxes. Not only did he save a lot of money on his taxes, but the consultant filed an amended form for the previous year and received a refund for that year. As the business progressed, Bill learned more and more the value of good tax planning.

## Introduction

The purpose of this chapter is to appraise the effect that **income taxes** and other withholdings have on wages and to introduce the concept of business income tax return preparation. (Note: Due to the ongoing modifications in the tax environment, you are encouraged to utilize local and current tax forms for this chapter.)

When Troy entered the workforce, he failed to realize that there would be a considerable difference between his service station gross earnings and his take-home pay because of the added responsibility of income tax (Figure 11–1). Also, his involvement with Sprout About, the entrepreneurial horticulture business he and Megan started, necessitated a need to understand how to file a business tax return for his income tax and **self-employment tax** liabilities (Figure 11–2). Troy was legally required to file a return. Ms. Green felt that because her class encouraged students to earn income, she had the responsibility to teach students how to meet tax responsibilities.

## ■ DOCUMENTS FILED WITH EMPLOYER

Prior to Troy's job interview at the service station, Mr. Shoupsky, the owner, asked Troy to provide proof of identification and of work eligibility. Troy wanted to be prepared for the interview; so he asked Ms. Green about the specifics of this request.

**FIGURE 11–1**

Reviewing local, state, and federal tax requirements is a part of being in the workforce.

**FIGURE 11-2**

This young man's welding business is so successful he has to pay income taxes.

"*M*r. Shoupsky cannot pay you without resident documentation. Troy, you have to properly identify yourself and indicate that you can legally work in the United States," Ms. Green remarked.

Ms. Green noted that the **I-9** form is used as proof of eligibility to work in the United States. She added that a driver's license and **social security** number are usually sufficient to indicate that you are eligible for employment.

The **W-4** is another form that is completed prior to employment. It is used by the employee to indicate the number of exemptions claimed. **Exemptions** are filed for each person who relies for financial support on the person earning wages. Usually, young unmarried people, like Troy, include only themselves as exemptions. However, each child and a spouse can be included as exemptions. Parents include their children as exemptions through early

**FIGURE 11-3**

Teaching about taxes is the focus of this small group session.

adulthood if they are full-time students. The number of exemptions partly determines the employee's tax withholdings (Figure 11–3).

# ■ TAKE-HOME PAY

Employee tax withholdings are just a portion of the amount of money that employers deduct from employees' gross wages. The tax withholdings are used to pay employees' state and federal taxes as well as the Federal Insurance Contribution Act **(FICA)** tax, also known as the social security tax. Other deductions include benefit contributions such as health insurance and retirement. Employers send the deducted wages to all the appropriate recipients on each payday (Table 11–1). After all deductions are made from your gross salary, what is left is your **take-home pay.**

### TABLE 11–1   PAYROLL CALCULATION SHEET

**Income**

| | |
|---|---|
| Hours worked (10) × hourly rate ($5.00) = regular wages.......................................$ | 50.00 |
| Hours of overtime worked × overtime rate = overtime wages ...................................+ | |
|    Gross pay.................................................................= | $50.00 |

**Withholding**

| | |
|---|---|
| Federal income tax withholding (9%).............................................$ | 4.50 |
| State income tax withholding (0.9%)...............................................+ | .45 |
| FICA withholding (13%)..........................................................+ | 6.50 |
| Other withholding..............................................................+ | |
|    Total withholding...........................................................= | $11.45 |
| **Net Pay**.............................................**(Gross pay-total withholding)** $ | 38.55 |

*T*roy contemplated, "If I work at Mr. Shoupsky's service station 10 hours each week for $5/hour, I won't bring home $50?"

*M*s. Green answered, "That is correct. Mr. Shoupsky will deduct state and federal taxes and FICA taxes from your paycheck. I would guess that you will take home around $38. Let's do some calculations and find out exactly what your take-home pay will be."

The first items that Ms. Green and Troy needed were the federal and state income tax rates. It was determined that they were 9% and 0.9% respectively. The FICA tax rate was 13%. Ms. Green provides Troy with a payroll calculation sheet (Table 11–1).

Ms. Green informed Troy that he will receive a payroll statement with each paycheck (Table 11–2). The statement includes information about the current pay period and it updates the financial records for the year (Table 11–3).

### TABLE 11–2　PAYROLL STATEMENT: FIRST PAY PERIOD

| Hours | | Earnings | | | | |
|---|---|---|---|---|---|---|
| Regular | Overtime | Regular | Overtime | Bonus | Other | Gross Pay |
| 10 | | $5.00 | $ | $ | $ | $50.00 |
| **Deductions** | | | | | | |
| FICA | Federal Tax | State Tax | Health Insurance | Retirement | Other | Net Pay |
| $6.50 | $4.50 | $.45 | $ | $ | $ | $38.55 |
| **Year to Date** | | | | | | |
| Year | Regular | Overtime | Gross Pay | FICA | Federal Tax | State Tax |
| 19xx | $50.00 | | $50.00 | $6.50 | $4.50 | $.45 |

### TABLE 11–3　PAYROLL STATEMENT: SECOND PAY PERIOD

| Hours | | Earnings | | | | |
|---|---|---|---|---|---|---|
| Regular | Overtime | Regular | Overtime | Bonus | Other | Gross Pay |
| 10 | | $5.00 | $ | $ | $ | $50.00 |
| **Deductions** | | | | | | |
| FICA | Federal Tax | State Tax | Health Insurance | Retirement | Other | Net Pay |
| $6.50 | $4.50 | $.45 | $ | $ | $ | $38.55 |
| **Year to Date** | | | | | | |
| Year | Regular | Overtime | Gross Pay | FICA | Federal Tax | State Tax |
| 19xx | $100.00 | | $100.00 | $13.00 | $9.00 | $.90 |

## PERSONAL RECORDS

Ms. Green advised Troy to save his payroll statements or **pay stubs** in a personal records file for at least three years, along with the **W-2** form. The Internal Revenue Service **(IRS)** can **audit** your tax returns up to three years after you filed them. In special instances they can extend the audit period to six years.

> **"I** thought we talked about a W-4 form. Where does this W-2 form come from?" Troy wondered.

The W-2 form, completed by the employer after the end of the year, documents the totals of **gross pay,** tax withholdings, and deductions. Several copies of this form are provided to the employee, usually in January. These copies are to be attached to the federal and state income tax forms when tax returns are filed.

## ■ FILING INCOME TAX RETURNS

Ms. Green acknowledged that tax laws change frequently; so her long-term advice to Troy included the following points. Obtain current federal and state income tax forms and instructions. Determine your filing status according to current tax laws. The selection of forms depends on income circumstances such as the amount of income, type of income, and the number of dependents or exemptions. Tax **schedules** are based on exemptions and used to determine your **tax bracket.** The information found in payroll records, pay stubs, or payroll statements is summarized at the end of the year on a W-2. The W-2 includes federal, state, and local tax payments, as well as gross wages and other deductions.

> **M**egan commented, "This is fine for Troy's service station job with Mr. Shoupsky, but what about our business, Sprout About? How do we account for profits or losses from our business?"

## ■ BUSINESS TAXES

### BUSINESS TAXABLE INCOME

Keeping good financial records is a must for any business. Taxable business income includes the sale of items raised or produced, the sale of items purchased for resale less purchase price, and income from custom work performed for others.

**FIGURE 11-4**

**To be depreciable, an asset must meet three basic requirements.**

1. The useful life can be determined.
   For example: the service life of a truck can be determined, the life of an acre of land cannot, therefore land cannot be depreciated.
2. It must have a useful life of more than one year.
3. It is used for business purposes.
   If an asset is used for business and personal use, only the business portion of the cost may be eligible for depreciation.

## BUSINESS DEDUCTIONS

The main business deductions are costs for expendable, or noncapital, assets, commonly called "normal production expenses." **Depreciation** is another source of business deductions. Depreciation comes from assets that have a useful life of more than one year (review Chapter 2) (Figure 11–4).

Machinery or buildings used in the production process may be written off during their useful life. The current IRS rules for property classifications and depreciation schedules are indispensable when making business decisions. However, tax depreciation can be different from depreciation methods used to guide management decisions.

> "*D*id you just imply that we might have to keep two sets of financial records for our business, Sprout About—one for the business and one for the IRS?" Megan questioned.

## ■ TAX MANAGEMENT

Most businesses keep two sets of books or financial records because IRS depreciation schedules accelerate at a different rate than the useful life rate that a business uses when it makes management decisions. The purpose of the tax management records is to maximize after-tax income, not to minimize taxes. The goals are to arrange to have little or no taxable income and to continuously evaluate financial decisions in light of taxes. This is accomplished by managing income, expenses, purchases, and the sale of capital assets.

> "*I*f you own your own business, do you still have to pay social security tax?" Megan pondered.

## SELF-EMPLOYMENT TAX

Self-employed individuals still have a social security tax liability. The self-employment tax is paid by self-employed individuals in lieu of FICA taxes. This process allows entrepreneurs to be eligible for social security benefits.

# Summary

The ever changing rules associated with taxation require people to keep good financial records and to stay up-to-date with current tax laws and regulations. This advice is especially true for entrepreneurs. The most important point in this chapter is that sound financial management incorporates taxation as a part of the business decision-making process and that some decisions, such as those affecting depreciation, may warrant different strategies for the IRS than those used for the business.

Troy, Megan, and Ms. Green learned a great deal in the past ten chapters. They will continue their adventures as Troy expands the Sprout About business, Megan decides to pursue a biotechnology career, and Ms. Green continues answering questions about record keeping, financial management, and business analysis.

### *SUPERVISED EXPERIENCE*

Success often leads to greater challenges, as this young man found out when he started to expand his yard care business. His business was in such demand that he hired some of his friends to help with the work. This step led to increased income and higher self-employment taxes (Figure 11–5). In addition, he paid the employer's share of social security taxes for his

**FIGURE 11–5**

This young man's yard care business was so productive he hired friends and paid self-employment taxes.

friends. Although addressing all the tax regulations were overwhelming at first, the rewards of higher profits made it worthwhile. He was able to finance his own pickup truck and attend the local community college after high school.

## CAREER OVERVIEW

Careers that utilize agricultural, business and scientific experiences include (Figure 11–6):

High school diploma: Dude Ranch Guide; Garden Center Trainee

Associate degree: Soil Research Technician; Crop Insurance Adjuster

Bachelor's degree: Agricultural Loan Officer; Agricultural Journalist

Advanced degree: International Agricultural Economist; Rural Sociologist; Agricultural Tax Lawyer

## DISCUSSION QUESTIONS

1. Troy works for Shoupsky's Service Station. He works 20 hours per week and is paid $6/hour once a week.
   - Nine percent of his income is withheld for federal income tax.
   - For state income tax, 0.9% of his income is withheld.
   - Thirteen percent of his income is deducted for FICA.

**FIGURE 11-6**

Studying to be an agricultural tax lawyer, this young man focused his studies to enhance his career goal.

## TABLE 11-4   TROY'S PAYROLL CALCULATION SHEET

**Income**

Hours worked × hourly rate = regular wages . . . . . . . . . . . . . . . . . . . . . . . . . . . . . . . . . . . . . . . . $ _____

Hours of overtime worked × overtime rate = overtime wages . . . . . . . . . . . . . . . . . . . . . . . . . . . . . + _____

   Gross pay . . . . . . . . . . . . . . . . . . . . . . . . . . . . . . . . . . . . . . . . . . . . . . . . . . . . . . . . . . . . . . . = _____

**Withholding**

Federal income tax withholding . . . . . . . . . . . . . . . . . . . . . . . . . . . . . . . . . . . . . . . . . . . . . . . . . $ _____

State income tax withholding . . . . . . . . . . . . . . . . . . . . . . . . . . . . . . . . . . . . . . . . . . . . . . . . . . + _____

FICA withholding. . . . . . . . . . . . . . . . . . . . . . . . . . . . . . . . . . . . . . . . . . . . . . . . . . . . . . . . . . . + _____

Other withholding . . . . . . . . . . . . . . . . . . . . . . . . . . . . . . . . . . . . . . . . . . . . . . . . . . . . . . . . . . + _____

   Total withholding. . . . . . . . . . . . . . . . . . . . . . . . . . . . . . . . . . . . . . . . . . . . . . . . . . . . . . . . = _____

**Net Pay** . . . . . . . . . . . . . . . . . . . . . . . . . . . . . . . . . . . . . . . . . . . . . .(Gross pay-total withholding) $ _____

## TABLE 11-5   TROY'S PAYROLL STATEMENT

| Hours | | Earnings | | | | |
|---|---|---|---|---|---|---|
| Regular | Overtime | Regular | Overtime | Bonus | Other | Gross Pay |
| 10 | | $ | $ | $ | $ | $ |
| **Deductions** | | | | | | |
| FCA | Federal Tax | State Tax | Health Insurance | Retirement | Other | Net Pay |
| $ | $ | $ | $ | $ | $ | $ |
| **Year to Date** | | | | | | |
| Year | Regular | Overtime | Gross Pay | FICA | Federal tax | State Tax |
| | $ | | $ | $ | $ | $ |

Use the payroll calculation sheet (Table 11–4) to determine Troy's deductions and net income. Then complete the payroll statement (Table 11–5). The amount recorded under **Net Pay** indicates the amount he will receive in his first paycheck of the new year.

2. Distinguish between tax-deductible and non-tax–deductible business expenses.

3. Distinguish between a deductible expense and a capital purchase.

## SUGGESTED ACTIVITIES

1. With play money, pay students a gross wage. After they have received the cash, collect the taxes usually deducted from a paycheck (use a tax rate appropriate to the students' actual situation). Facilitate discussion on net and gross wages and the purpose of deductions and taxes.

2. Adapt activity 1 to include annual tax reporting. (This lesson is especially appropriate during the first part of April.) Give some students refunds and have other students make payments to the IRS. Take the opportunity to explain how people can choose the amount deducted from their paychecks. Explain how this step directly affects the amount of taxes that they pay or the amount of refund that they receive.

3. Hand out a tax form and ask students to complete it. As frustration rises, ask students if there is more information that they need to know before they can complete the form. Ask students to explain the rationale for filing income tax returns.

4. Have students fill out sample W-4 and resident documentation forms.

## GLOSSARY

**Audit**   Complete examination of financial records.

**Depreciation**   Decrease in value of business assets caused by wear and obsolescence.

**Exemptions**   Deductions from income tax responsibility.

**FICA**   Federal Insurance Contribution Act, the tax withheld from an employee's paycheck and matched by the employer; the amount is invested in a trust account by the Social Security Administration for a time when the employee no longer works. *Also known as* social security tax or retirement tax.

**Gross Pay**   Total amount of pay that is due an employee before any withholding.

**Income Tax**   Tax levied on wages and salaries to finance government activities.

**I-9**   A form that documents eligibility to work in the United States. Proof of eligibility is required of every person for any job.

**IRS**   Internal Revenue Service, the tax-collecting agency for the federal government.

**Net Pay**   Gross pay minus all withholding. *See also* take home pay.

**Pay Stub**   Portion of the paycheck that the employee must retain for three years. The pay stub includes the amount of gross pay, all withholdings and their amounts, and the amount of net pay. *Also known as* payroll statement.

**Schedule**   Income tax form that itemizes deductions and exemptions.

**Self-employment Tax**   Tax liability paid by people who are self-employed instead of paying FICA taxes.

**Social Security**   A trust account managed by the Social Security Administration for the purpose of saving money for people to use when they are no longer able to work. A person is identified with a nine-digit social security number, also called a tax identification number.

**Take-home Pay**   Amount a person receives after taxes and other deductions are withheld from his or her earnings. *See also* net income, net pay.

**Tax Bracket**   Bracket, based on income, that determines what percentage of income will be paid in taxes and which forms to file as an income tax return.

**W-2**  Form, completed by the employer, that documents the annual totals of gross pay and all tax withholding for an employee that was reported during the year.

**W-4**  Form, filed by the employee with the employer, that documents the number of exemptions claimed. The number of exemptions partly determines the employee's tax withholding.

# Management Information System

## Introduction

The purpose of this chapter is to enable you to record information and to manage finances through the use of a management information system (MIS). When students are involved in a Supervised Experience Program, they must be able to keep accurate records, both financial and non-financial. And in any career, students must be able to manage their records and finances and be in constant touch with their financial standings. This chapter gives students an opportunity to use all the financial forms explained in this book and to complete additional records on a real-life situation in which they are the decision-makers. The value of the management information system presented in this chapter is in the hands-on experience of using the records in connection with live-fire management information.

Many of the pages in this management information system are non-financial pages and therefore were not addressed in the preceding chapters. However, the introductory pages in this chapter offer instructions on how to address all the pages in the system.

## ■ INSTRUCTIONS FOR THE MANAGEMENT INFORMATION SYSTEM

### RECORD OF SUPERVISED EXPERIENCE PROGRAM [FORM 1]

The Record of Supervised Experience Program is used as a summary record of all supervised experiences, including entrepreneurship, placement, improvement, research, and exploratory (include with other). These sheets, designed to be carried over from one year to the next, are designed to be completed at the end of each year of instruction.

### EXPLORATORY SUPERVISED EXPERIENCES [FORMS 2-3]

#### Planning Sheet [Form 2]

The planning sheet is used each year to plan exploratory supervised experiences, which often include research projects. Exploratory supervised experiences are designed to increase student agricultural awareness and/or agricultural literacy. For example, if students lack agricultural experience, then exploratory supervised experiences can increase their agricultural experience level. Research projects may require additional sheets to address the scientific process and the experimental procedures. A new sheet should be used for each major area of interest.

#### Exploratory Supervised Experience Records [Form 3]

This form provides space for students to record information on the activities that they planned in Form 3. They are to identify the activity (title), describe the project, list the objectives of the activity, list the completed activities, and identify outcomes or accomplishments.

## PLACEMENT EXPERIENCES [FORMS 4–6]

### Placement Experience Training Agreement [Form 4]

Placement Experience Training Agreements are designed to clarify responsibilities of all parties before engaging in work experiences or jobs. Prior to initiating any placement experiences, students must obtain permission to conduct this type of experience from the appropriate agency (i.e., school board, advisory committee, administration, etc.). Four parties—student, employer, teacher, and parent/guardian—are involved with this agreement, and all their signatures should appear on the agreement.

### Placement Experience Plan [Form 5]

The Placement Experience Plan is meant to be used to plan, identify, and record experiences, competencies, and related school instruction while working at the job. A special effort should be made so that students obtain a variety of experiences when employed.

### Placement Experience Evaluation Form [Form 6]

This form is designed to be used by the person, usually the employer, who has the best knowledge of the student's performance while at the placement site. The form can be completed periodically, such as monthly, quarterly, or annually. It is designed to help students improve their work experience skills and attitudes. Therefore, more frequent constructive advice is preferred to a glossy overview at the end of the experience.

## SCHOOL-BASED AGREEMENT FORMS [FORMS 7–8]

Certain schools may find that conducting school-based supervised experiences is more appropriate for their situations. You have two forms to choose from. Form A, the more generic one, allows schools the flexibility to generate their own set of expectations. Form B is more specific about the contributions from the school, school district, FFA chapter, and student. Specific school-based projects include greenhouse production sales, fruit sales, holiday wreath sales, and other such projects.

## IMPROVEMENT ACTIVITIES [FORM 9]

Improvement activities continue to be an important supervised experience for many students. Use this form for both planning and recording improvement projects. A separate form should be used for each major project. Additional sheets that record detailed information and photographs of the project are also encouraged.

## ENTERPRISE BUDGET [FORM 10]

The enterprise budget is a planning tool (see Chapter 9), used to estimate potential profit or return. Use one form for each potential new enterprise. Estimated costs, both variable and fixed, are subtracted from estimated revenue or income. The estimated profit can be compared with other potential opportunities as students decide on possible enterprises.

## PARTIAL BUDGET [FORM 11]

The partial budget is used for financial planning when a change is proposed (see Chapter 9). A new form is required for each proposed change. Students must identify and record new or additional costs, current income that will be lost or reduced, new or additional income, and current costs that will be lost or reduced. A positive difference on the partial budget form indicates that the proposed change is profitable.

## CASH FLOW BUDGET [FORM 12]

The cash flow budget, like the previous two budgets, is a planning tool reflecting the student's best estimation of the flow of income (cash inflow) and expenses (cash outflow) in a given time period (month, quarter, or year, as selected by the student). It usually follows a logistical time frame for the enterprise or business being analyzed, such as production cycle, month, quarter, and year.

■ The beginning cash balance is the estimated cash on hand at the beginning of the time period.

■ Cash inflows include all potential sources of revenue for the given time period.

■ Cash outflows include all the potential sources of expense for the time period. The cash balance is simply the difference between Total Cash Inflows and Total Cash Outflows. In the calculation of cash balance, a negative difference indicates a need to borrow funds to meet the current period's financial obligations.

■ Loan repayments include all principal and interest the student intends to pay back during the period.

■ Ending cash balance is calculated by adding the cash balance and the borrowed funds needed, then subtracting the loan repayment from the sum.

■ The debt outstanding becomes the borrowed funds needed minus the loan repayment. Students can chart the debt outstanding for each time period and use the information to determine whether their enterprises increase or reduce debt.

## INVENTORY FORMS [FORMS 13–15]

The inventories begin the financial record keeping portion of the management information system, which is designed for single or multiple enterprises. Additional pages of all forms [Forms 13–31] are needed if students want to keep records by enterprise, in which case the following requirement is important: *To the right of the word "Enterprise," students must identify whether the sheet contains enterprise or total values.*

This distinction becomes important when students apply for FFA awards. This financial information is required in the new National FFA Applications.

In addition, some of the MIS financial forms are slightly more advanced than the forms found in the previous chapters. (The concepts are the same, but additional components are required for a record of finances.) For example, in Chapter 3 on inventory, students are asked to determine only if an asset is classified as depreciable or non-depreciable. In the MIS, students are asked to determine further whether the non-depreciable asset is current or non-current and to conduct a beginning and ending inventory.

### Non-depreciable Inventory [Form 13]

Inventories (see Chapter 3) help to determine the value or net worth of the business. Non-depreciable inventory items consist of assets that do not depreciate or that are not kept for over one year or that are not valued at over $100. The following rules should be applied.

1. Current non-depreciable inventory items are held less than a year.
2. Non-current non-depreciable inventory items are held longer than a year.

Students are asked to conduct a beginning and ending inventory. Beginning inventories are usually conducted on January 1. Ending inventories are usually conducted on December 31.

### Depreciable Inventories [Forms 14–15]

Beginning inventories (see Chapter 3) are usually conducted on January 1 or when a business begins (i.e., in the summer or fall for first-time students). Ending inventories are usually conducted on December 31. Students are given a choice between using the straight-line schedule or the market value schedule. Both methods have benefits, and a local decision is required. However, once a decision is made, it must be followed consistently. (Note: Raised livestock kept for breeding purposes should be inventoried here.) Students are asked to add the depreciation values on the ending balance sheet and transfer the total to the accrual income statement [Form 24].

## BALANCE SHEET STATEMENT [FORMS 16–17]

The balance sheet statements (see Chapter 3) ask for beginning and ending values. Many of the balance sheet values are transferred from the inventories. In addition, many of the values are to be transferred to other statements. Many students will not have many assets or liabilities; so a school-based project may be an appropriate avenue for providing experience with in-depth record keeping.

## RECORD OF INCOME AND EXPENSES CODE SHEET [FORM 18]

This is the most important page in the MIS. Each entry on the record of income and expenses pages must be identified with a code from this page. All of the summaries and analyses depend on accurate identification during data entry. Most entries fall into two sections.

■ Section 2, cash paid for operating activities
■ Section 3, cash received from operating activities

Students are encouraged to utilize the second-level (e.g., 2.1), third-level (e.g., 2.1.1), and fourth-level (e.g., 2.1.3.1) categories to further delineate financial entries. Students are also encouraged to create other subcategories in the 11 main areas if they need more precise records.

## RECORD OF INCOME AND EXPENSES [FORM 19]

The actual entry of data occurs in this section. It is absolutely essential that each item be identified with a category code from Form 18. If multiple enterprises are recorded, then one sheet should be utilized and identified as the total sheet.

## INCOME AND EXPENSE SUMMARIES [FORMS 20–22]

These three summaries provide much of the data used to calculate the financial ratios. Again, proper identification and coding of the financial entries are crucial to the accuracy of any summarized reporting. Note that third- and fourth-level coded values are needed for calculations on Form 21. (If this last statement does not make sense, reread the instructions on the code sheet. Detailed instructions are included for each summary.)

## INCOME STATEMENTS [FORMS 23–24]

Either the cash or the accrual form may be used. Revisit Chapter 5 for an expanded discussion on income statements. The revenue and expenses sections can be detailed or summarized depending on the student's wishes.

## STATEMENT OF CASH FLOWS WORKSHEET [FORM 25]

The values for this worksheet are transferred from the income and expense summary [Forms 20–21]. Only 7 of 11 coded categories are calculated on this page. The other four coded categories are included on Form 26. Extremely detailed instructions are included on the worksheet.

## STATEMENT OF CASH FLOWS [FORM 26]

All 11 coded categories are included on this form (see Chapter 6). These values are important ingredients for the financial worksheet on Form 29.

## CASH FLOW STATEMENT [FORM 27]

The cash flow statement is designed to help students determine their cash surpluses and deficiencies. Revisit Chapter 6 for clarification on cash flow statements. It is not one of the four main statements required for financial analysis, but it is an effective planning tool and should be attempted by high school students.

## STATEMENT OF OWNER EQUITY [FORM 28]

All of the values for this form are transferred in from other reports or calculated from the transferred data. Comparing and understanding the differences between the ending owner equity on Form 28 and the ending owner equity on Form 17 is challenging, but important.

## FINANCIAL WORKSHEET [FORM 29]

All of the values needed to calculate the 16 financial ratios are assembled here. Specific instructions are provided for each item. Proper coding and careful calculating prior to completing this form are essential for accurate values.

## FINANCIAL RATIOS [FORM 30]

These two pages are the heart of financial management. If students are able to calculate most of the ratios, then they will be able to make informed decisions about their business. The calculations are relatively simple if the previous MIS pages are accurate. Reviewing Chapter 8 becomes a must when interpreting the financial values and ratios.

## TREND ANALYSES [FORM 31]

This longitudinal chart is designed to assess the business over a period of time. Students should be encouraged to chart their progress on at least some of the items listed. A simple transfer of values from Form 30 is required.

## RESUME [FORMS 32–34]

The entries on these three pages provide a record for students as they compile information for their resumes, which are required on the new FFA awards.

## THE DIARY [FORM 35]

The entries on this page are intended to provide students with information when writing the story section in the new FFA award applications.

## SUPERVISION OF STUDENT'S EXPERIENCE PROGRAM [FORM 36]

On this page, the instructor can record grades on the student's MIS.

**FORM 1    RECORD OF SUPERVISED EXPERIENCE PROGRAM\***

| Student Name | School Years: _____ To _____ |
|---|---|
| Address | Phone No. |
| Name of Parent/Guardian | Work Phone No. |

**Proficiency Area Interests**
**(identify areas from the latest FFA list of proficiency areas)**

Entrepreneurship Areas:

_____
_____
_____
_____
_____

Placement Areas:

_____
_____
_____
_____
_____

| | First Year _____ to _____ | | Second Year _____ to _____ | |
|---|---|---|---|---|
| | **Project** | **Scope** | **Project** | **Scope** |
| entrepreneurship and placement projects and scope | | | | |
| | | | | |
| | | | | |
| | | | | |
| | | | | |
| improvement projects and scope | | | | |
| | | | | |
| | | | | |
| | | | | |
| | | | | |
| research and other projects and scope | | | | |
| | | | | |
| | | | | |
| | | | | |
| | | | | |
| Instructor's Comments | | | | |
| Instructor's Signature | | | | |

*Source: The National Council for Agricultural Education, DECISIONS & DOLLARS, 1995

**FORM 1    RECORD OF SUPERVISED EXPERIENCE PROGRAM (CONTINUED)**

| Student Name | | | School Years: _____ To _____ | |
|---|---|---|---|---|
| | **Third Year _____ to _____** | | **Fourth Year _____ to _____** | |
| | **Project** | **Scope** | **Project** | **Scope** |
| entrepreneurship and placement projects and scope | | | | |
| | | | | |
| | | | | |
| | | | | |
| | | | | |
| improvement projects and scope | | | | |
| | | | | |
| | | | | |
| | | | | |
| | | | | |
| research and other projects and scope | | | | |
| | | | | |
| | | | | |
| | | | | |
| | | | | |
| Instructor's Comments | | | | |
| Instructor's Signature | | | | |

**Additional Information and Record of Visits**

| Date | |
|---|---|
| | |
| | |
| | |
| | |
| | |
| | |
| | |

**FORM 2   EXPLORATORY SUPERVISED EXPERIENCE PLANNING SHEET***

Student Name:_____

Length of plan: _____  starting date: _____  ending date: _____

Agricultural interest areas:

Career goals:

| Planned experiences: | Assistance needed: | Evaluation: | Date: |
|---|---|---|---|
| | | | |

Student signature:

Teacher signature:

Parent/Guardian signature:

*Source: The National Council for Agricultural Education, DECISIONS & DOLLARS, 1995

**FORM 3    EXPLORATORY SUPERVISED EXPERIENCE RECORDS\***

| Title: |
| --- |
| Overall description: |
| Objectives: |

| Specific activities completed: | Outcomes: (What did you accomplish?) |
| --- | --- |
| Date completed: | Grade: |

\*Source: The National Council for Agricultural Education, DECISIONS & DOLLARS, 1995

## FORM 4   PLACEMENT EXPERIENCE TRAINING AGREEMENT[1]*

Paid _____
Non-paid_____

To provide a basis of understanding and to promote sound business relationships, this agreement is established on

_____, _____. Placement will begin on _____, _____ and will end on or about _____, _____, unless the

arrangement becomes unsatisfactory to either party.

Name of agribusiness/farm/school/community facility _____

Name of supervisor _____

*The usual placement hours will be as follows:*

While attending school _____    When not attending school _____

Provision for overtime _____

Liability insurance coverage (type and amount) _____

Length of trial period and wage rate _____

Wage rate for the remainder of the agreement period _____

Frequency of payment _____

### It is understood that the employer will:

Provide the student with opportunities to learn how to do as many jobs as possible, with particular reference to those contained in the Placement Plans.

Instruct the student in ways of doing his/her work and handling his/her management problems.

Help the teacher make an honest appraisal of the student's performance.

Avoid subjecting the student to unnecessary hazards.

Notify the parent/guardian and teacher immediately in case of accident or sickness and if any other serious problem arises.

Assign the student new responsibilities in keeping with his/her progress.

Cooperate with the teacher in arranging a conference with the teacher on supervisory visits.

Other (Include all other points, on a separate sheet, that the employer will have the responsibility to provide.).

### The student agrees to:

Do an honest day's work.

Keep the employer's interest in mind; be punctual, dependable, and loyal.

Follow instructions, avoid unsafe acts, and be alert to unsafe conditions.

Be courteous and considerate of the employer, the employer's family, and other employees.

Keep records of occupational experiences and make required reports.

Achieve competencies indicated in the placement plan.

Other (In case there are additional student responsibilities, they should be included on a separate sheet.).

[1]States or schools which have placement agreements may insert their own agreement.
*Source: The National Council for Agricultural Education, DECISIONS & DOLLARS, 1995

## FORM 4    PLACEMENT EXPERIENCE TRAINING AGREEMENT (CONTINUED)

**The teacher, on behalf of the school, agrees to:**

Check and approve the placement center.

Provide a copy of the agreement to the school and employer.

Schedule class instruction to prepare students for occupational experience.

Visit the student on the job at frequent intervals for the purpose of instruction and to insure that the student gets the most education out of the experience.

Show discretion at the time and circumstances of these visits, especially when the work is pressing.

Assist the student in obtaining a work permit and developing a work plan.

Other (Additional teacher responsibilities should be added on a separate sheet.).

**The parents/guardians agree to:**

Assist in promoting the value of the student's experience by cooperating with the employer and teacher.

Satisfy themselves in regard to the living and working conditions made available to the student.

Assist in providing transportation to and from the placement center according to work schedule.

Other (Additional items agreed to by the parents/guardians should be included on a separate sheet.).

**All parties agree to:** An initial trial period of _____ working days to allow the student to adjust to the job. Discuss the issues with the teacher before ending the placement experience.

Other (List on a separate sheet.).

**Student** _____ (signature)

Address _____

Social Security No. _____ Telephone No. _____

Date of Birth _____ Age _____

**Parent/Guardian** _____ (signature)

Address _____ Telephone No. _____

**Employer** _____ (signature)

Address _____ Telephone No. _____

**Teacher** _____ (signature)

Address _____

Home Telephone No. _____ School Telephone No. _____

## FORM 5 PLACEMENT EXPERIENCE PLAN*

Name of student: _____ Teacher: _____

Student's career/professional objective: _____

Beginning date: _____ Ending date: _____

Supervised experience training station: _____

Paid: _____ Non-paid: _____

| Experiences/Competencies | Date Accomplished | School-Related Instruction | ✓ When Done |
|---|---|---|---|
| | | | |
| | | | |

*Source: The National Council for Agricultural Education, Decisions & Dollars, 1995

## FORM 6    PLACEMENT EXPERIENCE EVALUATION FORM*

Name of student: _____    Placement station: _____

Beginning date: _____    Ending date: _____

Number of days late: _____    Number of days absent: _____

**Directions:** On the items below, please rate the student who is under your auspices as part of the Placement Experience program. Rate the student by circling the best numerical descriptor that corresponds, in your opinion, to the quality of the student's on-the-site behavior.

Rating Scale:  4 = superior;  3 = above average;  2 = average;  1 = below average;  0 = unsatisfactory

|  | (circle one) |  | (circle one) |
|---|---|---|---|
| 1. Attitude toward job | 4 3 2 1 0 | 7. Accepts responsibility | 4 3 2 1 0 |
| 2. Works cooperatively with others | 4 3 2 1 0 | 8. Applies knowledge gained | 4 3 2 1 0 |
| 3. Follows directions | 4 3 2 1 0 | 9. Exercises good judgement | 4 3 2 1 0 |
| 4. Shows respect for tools, materials, and equipment | 4 3 2 1 0 | 10. Accepts constructive criticism | 4 3 2 1 0 |
| 5. Uses time efficiently | 4 3 2 1 0 | 11. Displays good conduct, grooming, and discipline | 4 3 2 1 0 |
| 6. Performs assignments satisfactorily | 4 3 2 1 0 | 12. Shows an understanding and use of safety procedures | 4 3 2 1 0 |

I.   Please comment on new skills the student has learned or knowledge gained:

II.  General comments:

III. If you had to assign a letter grade (A, B, C, D, or F) on this student's report card, what letter grade would it be?

IV. How could this training station be improved to better serve the needs of the (A) student and (B) the owner/operator?

A.

B.

*(please use the back of this sheet for additional writing space)*

**Evaluator** _____ (signature & date)

*Source: The National Council for Agricultural Education, Decisions & Dollars, 1995

## FORM 7 · SCHOOL-BASED AGREEMENT FORM A*

Name of student: _____

Name and short description of project: _____

_____

Names of other students on the project: _____

_____

### Provisions & Benefits

| School agrees to provide: | School shall receive: |
|---|---|
| | |
| **FFA chapter agrees to provide:** | **FFA chapter shall receive:** |
| | |
| **Student agrees to provide:** | **Student shall receive:** |
| | |

**Student** _____ (signature)

Address _____

Social Security No. _____ Telephone No. _____

**Parent/Guardian** _____ (signature)

Address _____ Telephone No. _____

**School Official** _____ (signature)

Address _____ Telephone No. _____

**Teacher** _____ (signature)

Address _____

Home Telephone No. _____ School Telephone No. _____

*Source: The National Council for Agricultural Education, DECISIONS & DOLLARS, 1995

## FORM 8   SCHOOL-BASED AGREEMENT FORM B*

Name of student: _____

Name and short description of project: _____

_____

Names of other students on the project: _____

_____

### Provisions & Benefits

| School agrees to provide/pay: | School shall receive: |
|---|---|
| _____ % of all enterprise operating expenses<br>_____ % of all enterprise overhead expenses<br>_____ % of other items described here: | _____ % of the enterprise or its value<br>_____ % of the enterprise income<br>_____ % of other items described here: |
| **FFA chapter agrees to provide/pay:** | **FFA chapter shall receive:** |
| _____ % of all enterprise operating expenses<br>_____ % of all enterprise overhead expenses<br>_____ % of other items described here: | _____ % of the enterprise or its value<br>_____ % of the enterprise income<br>_____ % of other items described here: |
| **Student agrees to provide/pay:** | **Student shall receive:** |
| _____ % of all enterprise operating expenses<br>_____ % of all enterprise overhead expenses<br>_____ % of other items described here:<br>(circle one)<br>_____ hours of work per day/week/month | _____ % of the enterprise or its value<br>_____ % of the enterprise income<br>_____ % of other items described here: |

**Student** agrees to provide project plans, records, and reports as required:

_____ (signature)

Address _____

Social Security No. _____ Telephone No. _____

**Parent/Guardian** agrees to discuss any issues concerning the project with the teacher:

_____ (signature)

Address _____ Telephone No. _____

**School Official** agrees that the project has educational value and is an integral part of the program:

_____ (signature)

Address _____ Telephone No. _____

**Teacher** agrees to evaluate the student's work habits and competencies gained:

_____ (signature)

Address _____

Home Telephone No. _____ School Telephone No. _____

*Source: The National Council for Agricultural Education, DECISIONS & DOLLARS, 1995

## FORM 9  IMPROVEMENT ACTIVITIES*

Use this form to plan and record results of one of your improvement activities. The scope of the project should be broad enough to include the planning and developing of many skills. It should involve responsibilities over a period of time.

Name of Student

|  | Plans/Estimated | Results/Actual |
|---|---|---|
| Kind of improvement: |  |  |
| Size: |  |  |
| Things to improve: | 1.<br>2.<br>3. | 1.<br>2.<br>3. |
| COSTS:<br>Hours of not hired labor: | hours | hours |
| Hired labor costs: | $ | $ |
| Supplies: | 1. $<br>2. $<br>3. $ | 1. $<br>2. $<br>3. $ |
| Equipment rental: | $ | $ |
| Total costs: | $ | $ |
| Results: | 1.<br>2.<br>3. | 1.<br>2.<br>3. |

Starting date:                    Closing date:

Teacher comments and signature:

*Source: The National Council for Agricultural Education, DECISIONS & DOLLARS, 1995

**FORM 10    ENTERPRISE BUDGET***

| Items | Value/unit | Item totals | Total value |
|---|---|---|---|
| Income: | | | |
| | | Total income (A) | $ |
| Variable costs: | | | |
| | | Total variable costs (B) | $ |
| Fixed costs: | | | |
| | | Total fixed costs (C) | $ |
| Profit or return: | | (A − B − C = D) | $ |

*Source: The National Council for Agricultural Education, Decisions & Dollars, 1995

**FORM 11    PARTIAL BUDGET***

| ADDITIONAL COSTS: | | |
|---|---|---|
| Item | Value | |
| | $ | |
| | | |
| | | |
| | | |
| | | |
| | | |
| | | |
| Subtotal | $ | |

| ADDITIONAL INCOME: | | |
|---|---|---|
| Item | Value | |
| | $ | |
| | | |
| | | |
| | | |
| | | |
| | | |
| | | |
| Subtotal | $ | |

| REDUCED INCOME: | | |
|---|---|---|
| Item | Value | |
| | $ | |
| | | |
| | | |
| | | |
| | | |
| | | |
| Subtotal | $ | |

| REDUCED COSTS: | | |
|---|---|---|
| Item | Value | |
| | $ | |
| | | |
| | | |
| | | |
| | | |
| | | |
| Subtotal | $ | |

| A. Total annual additional costs and reduced income: | $ |
|---|---|

| B. Total annual additional income and reduced costs: | $ |
|---|---|

| NET CHANGE IN PROFIT    (B minus A) | $ |
|---|---|

*Source: The National Council for Agricultural Education, DECISIONS & DOLLARS, 1995

**FORM 12    CASH FLOW BUDGET***

| Time Periods | 1 | 2 | 3 | 4 |
|---|---|---|---|---|
| Beginning Cash Balance | $ | $ | $ | $ |
| Cash Inflow: | | | | |
| product sales . . . . . . . . . . . . . . . . . . . . . . . . . . . . . . | | | | |
| capital sales . . . . . . . . . . . . . . . . . . . . . . . . . . . . . . . | | | | |
| miscellaneous cash revenue . . . . . . . . . . . . . . . . . . | | | | |
| Total Cash Inflow (A) | $ | $ | $ | $ |
| Cash Outflow: operating expenses . . . . . . . . . . . . . . . . . . . . . . . . . | $ | $ | $ | $ |
| capital purchases . . . . . . . . . . . . . . . . . . . . . . . . . . . | | | | |
| miscellaneous expenses . . . . . . . . . . . . . . . . . . . . . | | | | |
| Total Cash Outflow (B) | $ | $ | $ | $ |
| Cash Balance (A − B = C) | $ | $ | $ | $ |
| Borrowed funds needed (D) [Needed if "C" is negative] | $ | $ | $ | $ |
| Loan repayments (E) [principal and interest] | $ | $ | $ | $ |
| Ending Cash Balance (C + D − E = F) | $ | $ | $ | $ |
| Debt Outstanding (D − E = G) | $ | $ | $ | $ |

*Source: The National Council for Agricultural Education, DECISIONS & DOLLARS, 1995

## FORM 13    NON-DEPRECIABLE INVENTORY*

**Enterprise:**

### Current Non-Depreciable Inventory

Beginning Values: Date

Ending Values: Date

| Item | Total Units | Market Value/Unit | Total Market Value | Item | Total Units | Market Value/Unit | Total Market Value |
|------|------------|-------------------|--------------------|------|------------|-------------------|--------------------|
|  |  |  |  |  |  |  |  |
|  |  |  |  |  |  |  |  |
|  |  |  |  |  |  |  |  |
|  |  |  |  |  |  |  |  |
|  |  |  |  |  |  |  |  |
|  |  |  |  |  |  |  |  |
|  |  |  |  |  |  |  |  |
|  |  |  |  |  |  |  |  |
|  |  |  |  |  |  |  |  |
|  |  | TOTAL | $    (1) |  |  | TOTAL | $    (2) |

### Non-Current Non-Depreciable Inventory

Beginning Values: Date

Ending Values: Date

| Item | Total Units | Market Value/Unit | Total Market Value | Item | Total Units | Market Value/Unit | Total Market Value |
|------|------------|-------------------|--------------------|------|------------|-------------------|--------------------|
|  |  |  |  |  |  |  |  |
|  |  |  |  |  |  |  |  |
|  |  |  |  |  |  |  |  |
|  |  |  |  |  |  |  |  |
|  |  |  |  |  |  |  |  |
|  |  |  |  |  |  |  |  |
|  |  |  |  |  |  |  |  |
|  |  | TOTAL | $    (3) |  |  | TOTAL | $    (4) |

Transfer totals (1), (2), (3), and (4) to the balance sheets **[Forms 16 & 17]**.
*Source: The National Council for Agricultural Education, DECISIONS & DOLLARS, 1995

**FORM 14   DEPRECIABLE INVENTORY\*\***

**Enterprise:**     **Beginning Values**     **Date**

| Item | Date Acquired | Total Units (a) | Cost/Unit (b) | Total Acquisition Cost (a × b = c) | Depreciation (d) | Total Depreciation Value (c − d = e) | Current Market Value/Unit (f) | Total Market Value (a × f = g) |
|---|---|---|---|---|---|---|---|---|
| Breeding Livestock\* | | | Straight-Line Schedule | | | | Market Schedule | |
| | | | | | | | | |
| | | | | | | | | |
| | | | | | | | | |
| | | | | | | | | |
| | | | | | | | | |
| | | | | | | | | |
| | | | | | | | | |
| | | | | | Subtotal (A) | $ | Subtotal | $ |
| Machinery and Equipment | | | Straight-Line Schedule | | | | Market Schedule | |
| | | | | | | | | |
| | | | | | | | | |
| | | | | | | | | |
| | | | | | | | | |
| | | | | | | | | |
| | | | | | Subtotal (B) | $ | Subtotal | $ |
| | | | | | TOTAL (A + B = C) | $   (5) | TOTAL | $   (5) |
| Real Estate and Improvements | | | Straight-Line Schedule | | | | Market Schedule | |
| | | | | | | | | |
| | | | | | | | | |
| | | | | | | | | |
| | | | | | | | | |
| | | | | | TOTAL (D) | $   (6) | TOTAL | $   (6) |

\*When raised livestock reaches maturity include them here. If you use the straight-line schedule, include the year of maturity as date acquired; complete columns: (a); (b)—treat this as the current market value/unit for the year they were added to the depreciable inventory; (c); (d) & (e). Select either the straight-line or market schedule and transfer totals (5) and (6) to the balance sheet **[Form 16]**.

\*\*Source: The National Council for Agricultural Education, DECISIONS & DOLLARS, 1995

**FORM 15    DEPRECIABLE INVENTORY\*\***

Enterprise:                              **Ending Values**                                        **Date**

| Item | Date Acquired | Total Units (a) | Cost/Unit (b) | Total Acquisition Cost (a × b = c) | Depreciation (d) | Total Depreciation Value (c − d = e) | Current Market Value/Unit (f) | Total Market Value (a × f = g) |
|---|---|---|---|---|---|---|---|---|
| Breeding Livestock\* | | | | Straight-Line Schedule | | | Market Schedule | |
| | | | | | | | | |
| | | | | | | | | |
| | | | | | | | | |
| | | | | | | | | |
| | | | | | | | | |
| | | | | | | | | |
| | | | | | | Subtotal (A) | $ | Subtotal | $ |
| Machinery and Equipment | | | | Straight-Line Schedule | | | Market Schedule | |
| | | | | | | | | |
| | | | | | | | | |
| | | | | | | | | |
| | | | | | | | | |
| | | | | | | Subtotal (B) | $ | Subtotal | $ |
| | | | | | TOTAL (A + B = C) | $   (7) | TOTAL | $   (7) |
| Real Estate and Improvements | | | | Straight-Line Schedule | | | Market Schedule | |
| | | | | | | | | |
| | | | | | | | | |
| | | | | | | | | |
| | | | | | | | | |
| | | | | | TOTAL (D) | $   (8) | TOTAL | $   (8) |

\*When raised livestock reaches maturity include them here. If you use the straight-line schedule, include the year of maturity as date acquired; complete columns: (a); (b)—treat this as the current market value/unit for the year they were added to the depreciable inventory; (c); (d) & (e).

Select either the straight-line or market schedule and transfer totals (7) and (8) to the balance sheet **[Form 17]**.

Add values in the Depreciation column (d) and transfer total to (27) income statements **[Form 23 or 24]**.

\*\*Source: The National Council for Agricultural Education, DECISIONS & DOLLARS, 1995

**FORM 16    BALANCE SHEET STATEMENT[1]***

Enterprise:　　　　　　**Beginning Values**　　　　　　Date

| Assets | | | Liabilities and Owner Equity | | |
|---|---|---|---|---|---|
| Current assets | Value | | Current liabilities | Value | |
| cash on-hand, checking & savings | | (15) | accounts & notes payable | | |
| cash value—bonds, stocks, life insurance | | | current portion of non-current debt[1] | | |
| notes & accounts receivable | | | accrued interest on current liabilities | | (32a) |
| non-depreciable inventory | | (1) | accrued interest on non-current liabilities | | (16a) |
| other | | | other | | |
| Total current assets (A) | $ | | Total current liabilities (D) | $ | |
| Non-current assets | Value | | Non-current liabilities | Value | |
| non-depreciable inventory | | (3) | notes & chattel mortgages (total debt minus current portion)[1] | | |
| depreciable inventory | | (5) | real estate mortgages/contracts (total minus current portion) | | |
| real estate & improvements | | (6) | other | | |
| other | | | | | |
| Total non-current assets (B) | $ | | Total non-current liabilities (E) | $ | |
| | | | Total liabilities (D + E = F) | $ | (10) |
| | | | Owner equity (C − F = G) | $ | (11) |
| Total assets (A + B = C) | $ | (9) | Total liabilities and owner equity (F + G = H) | $ | |
| Liquidity measures: #1 working capital (A − D) | | | #2 current ratio (A ÷ D) | | |
| Solvency measures: #4 debt-to-asset ratio (F ÷ C) | | | #5 equity-to-asset ratio (G ÷ C) | | |
| #6 debt-to-equity ratio (F ÷ G) | | | | | |

Transfer values [1], [3], [5], and [6] from inventories [**Forms 13 & 14**].
Transfer value [11] to ending balance sheet [**Form 17**] and to SOE [**Form 28**].
Transfer values [16a] and [32a] to income and expense summary [**Form 21**].
Transfer value [15] to statement of cash flow [**Form 26**] and financial worksheet [**Form 29**].
Transfer values [9], [10], and [11] to financial worksheet [**Form 29**].
[1]The current portion of non-current debt when added to the notes and chattel mortgages (total minus current portion) should equal the total debt.
*Source: The National Council for Agricultural Education, DECISIONS & DOLLARS, 1995

## FORM 17  BALANCE SHEET STATEMENT[1]*

**Enterprise:**          **Ending Values**                    **Date**

| Assets | | | Liabilities and Owner Equity | | |
|---|---|---|---|---|---|
| Current assets | Value | | Current liabilities | Value | |
| cash on-hand, checking & savings | | | accounts & notes payable | | |
| cash value—bonds, stocks, life insurance | | | current portion of non-current debt[1] | | |
| notes & accounts receivable | | | accrued interest on current liabilities | | (32b) |
| non-depreciable inventory | (2) | | accrued interest on non-current liabilities | | (16b) |
| other | | | other | | |
| Total current assets (A) | $ | (17) | Total current liabilities (D) | $ | (18) |
| Non-current assets | Value | | Non-current liabilities | Value | |
| non-depreciable inventory | (4) | | notes & chattel mortgages (total debt minus current portion)[1] | | |
| depreciable inventory | (7) | | real estate mortgages/contracts (total minus current portion) | | |
| real estate & improvements | (8) | | other | | |
| other | | | | | |
| Total non-current assets (B) | $ | | Total non-current liabilities (E) | $ | |
| | | | Total liabilities (D + E = F) | $ | (13) |
| | | | Owner equity (C − F = G) | $ | (14) |
| Total assets (A + B = C) | $ | (12) | Total liabilities and owner equity (F + G = H) | $ | |
| Gain or loss in owner's equity ($ (14) − $ (11) = I) $ | | | | | |
| Liquidity measures: # 1 working capital (A − D) | | | # 2 current ratio (A ÷ D) | | |
| Solvency measures: # 4 debt-to-asset ratio (F ÷ C) | | | # 5 equity-to-asset ratio (G ÷ C) | | |
| # 6 debt-to-equity ratio (F ÷ G) | | | | | |

Transfer values (2), (4), (7), and (8) from inventories **[Forms 13 & 15]**.
Transfer value (11) from beginning balance sheet **[Form 16]**.
Transfer values (16b) and (32b) to income and expense summary **[Form 21]**.
Transfer values (12), (13), (14), (17), and (18) to financial worksheet **[Form 29]**.
[1]The current portion of non-current debt when added to the notes and chattel mortgages (total minus current portion) should equal the total debt.
*Source: The National Council for Agricultural Education, DECISIONS & DOLLARS, 1995

**FORM 18    RECORD OF INCOME AND EXPENSES CODE SHEET\***

The following items are coded and the code number should be included in the code column on the following "Record on Income and Expenses" pages. You may include additional codes if you desire.

1 cash withdrawals for family living:
2 cash paid for operating activities:
  2.1 cash paid for operating activities
    2.1.1 feeder animal
    2.1.2 feed purchases
    2.1.3 operating expenses—note you may want to further categorize this item such as: 2.1.3.1—fuel; 2.1.3.2—oil/lubricants; 2.1.3.3—fertilizer; etc.
    2.1.4 interest expenses—paid in cash or by loan renewal
      2.1.4.1 operating loans & accounts payable[33]
      2.1.4.2 term debt & capital leases[22]
    2.1.5 hedging account deposits
  2.2 cash expenses paid in non business operations
  2.3 income and social security taxes paid in cash
  2.4 extraordinary items paid in cash
  2.5 cash withdrawal for family living
  2.6 cash withdrawals for investment into personal assets
3 cash received from operating activities:
  3.1 cash received from business operations (all cash sales from income statement)
    3.1.1 feeder livestock/poultry sales
    3.1.2 crops/feed
    3.1.3 custom work
    3.1.4 gov't payments
  3.2 cash received from non business income and operations
    3.2.1 wages
    3.2.2 interest
  3.3 extraordinary items received in cash
4 cash paid for investing activities:
  4.1 breeding & dairy livestock
  4.2 machinery & equipment

  4.3 business real estate; other business assets
  4.4 capital leased assets
  4.5 bonds & securities; investments in other entities; other non business assets
5 cash received from investing activities:
  5.1 raised breeding & dairy livestock (not capitalized & not depreciated)
  5.2 purchased & raised breeding & dairy livestock (capitalized & depreciated)
  5.3 machinery & equipment
  5.4 business real estate; other business assets
  5.5 bonds & securities; investments in other entities; other non business assets
6 distribution of dividends, capital, or gifts:
  6.1 cash portion
  6.2 property portion
7 cash received from equity contributions:
  7.1 gifts & inheritances[41]
    7.1.1 cash portion[41a]
    7.1.2 property portion[41b]
  7.2 additions to paid-in capital including investments of personal assets into the business[42]
    7.2.1 cash portion[42a]
    7.2.2 property portion[42b]
  7.3 investments of personal assets into the business
8 principal paid on term debt & capital leases:
  8.1 term debt principal payments: scheduled payments
  8.2 term debt principal payments: unscheduled payments
  8.3 principal portion of payments on capital leases
9 operating loans received (including interest paid by loan renewal):
10 proceeds from term debt (term debt financing; loans received):
11 operating debt principal payments:

Note: It is important to record noncash items (such as: receiving pasture in exchange for labor, no cash was involved, but a value should be established on both pasture & labor). Include them as you would cash items, place a value on the items, and identify them with a code. You may want to separate cash and noncash items during your analysis.
Note: values [22], [33], [41], and [42] are calculated on **[Form 21]**.
*Source: The National Council for Agricultural Education, DECISIONS & DOLLARS, 1995

**FORM 19    RECORD OF INCOME AND EXPENSES\*\***

**Enterprise:**

| Date | Item (including labor/hours) | Code* | Income | Expenses | Balance |
|------|------------------------------|-------|--------|----------|---------|
| | Beginning Balance | | | | |
| | | | | | |
| | | | | | |
| | | | | | |
| | | | | | |
| | | | | | |
| | | | | | |
| | | | | | |
| | | | | | |
| | | | | | |
| | | | | | |
| | | | | | |
| | | | | | |
| | | | | | |
| | | | | | |
| | | | | | |
| | | | | | |
| | | | | | |
| | | | | | |
| | | | | | |
| | | | | | |
| | | | | | |
| | Ending Balance | | | | |

\*Indicate category code from **[Form 18]**.

## FORM 19   RECORD OF INCOME AND EXPENSES** *(CONTINUED)*

**Enterprise:**

| Date | Item (including labor/hours) | Code* | Income | Expenses | Balance |
|------|------------------------------|-------|--------|----------|---------|
| Beginning Balance | | | | | |
| | | | | | |
| | | | | | |
| | | | | | |
| | | | | | |
| | | | | | |
| | | | | | |
| | | | | | |
| | | | | | |
| | | | | | |
| | | | | | |
| | | | | | |
| | | | | | |
| | | | | | |
| | | | | | |
| | | | | | |
| | | | | | |
| | | | | | |
| | | | | | |
| | | | | | |
| | | | | | |
| | | | | | |
| | | | | | |
| Ending Balance | | | | | |

*Indicate category code from **[Form 18]**.

*(continued)*

## FORM 19   RECORD OF INCOME AND EXPENSES** *(CONTINUED)*

**Enterprise:**

| Date | Item (including labor/hours) | Code* | Income | Expenses | Balance |
|------|------------------------------|-------|--------|----------|---------|
| | Beginning Balance | | | | |
| | | | | | |
| | | | | | |
| | | | | | |
| | | | | | |
| | | | | | |
| | | | | | |
| | | | | | |
| | | | | | |
| | | | | | |
| | | | | | |
| | | | | | |
| | | | | | |
| | | | | | |
| | | | | | |
| | | | | | |
| | | | | | |
| | | | | | |
| | | | | | |
| | | | | | |
| | | | | | |
| | | | | | |
| | | | | | |
| | Ending Balance | | | | |

*Indicate category code from **[Form 18]**.

## FORM 19    RECORD OF INCOME AND EXPENSES** (CONTINUED)

**Enterprise:**

| Date | Item (including labor/hours) | Code* | Income | Expenses | Balance |
|------|------------------------------|-------|--------|----------|---------|
| Beginning Balance | | | | | |
| | | | | | |
| | | | | | |
| | | | | | |
| | | | | | |
| | | | | | |
| | | | | | |
| | | | | | |
| | | | | | |
| | | | | | |
| | | | | | |
| | | | | | |
| | | | | | |
| | | | | | |
| | | | | | |
| | | | | | |
| | | | | | |
| | | | | | |
| | | | | | |
| | | | | | |
| | | | | | |
| Ending Balance | | | | | |

*Indicate category code from **[Form 18]**.

*(continued)*

**FORM 19    RECORD OF INCOME AND EXPENSES\*\* (CONTINUED)**

**Enterprise:**

| Date | Item (including labor/hours) | Code* | Income | Expenses | Balance |
|------|------------------------------|-------|--------|----------|---------|
| | Beginning Balance | | | | |
| | | | | | |
| | | | | | |
| | | | | | |
| | | | | | |
| | | | | | |
| | | | | | |
| | | | | | |
| | | | | | |
| | | | | | |
| | | | | | |
| | | | | | |
| | | | | | |
| | | | | | |
| | | | | | |
| | | | | | |
| | | | | | |
| | | | | | |
| | | | | | |
| | | | | | |
| | | | | | |
| | | | | | |
| | | | | | |
| | | | | | |
| | | | | | |
| | | | | | |
| | Ending Balance | | | | |

*Indicate category code from [Form 18].

**FORM 19    RECORD OF INCOME AND EXPENSES** *(CONTINUED)*

**Enterprise:**

| Date | Item (including labor/hours) | Code* | Income | Expenses | Balance |
|------|------------------------------|-------|--------|----------|---------|
| Beginning Balance | | | | | |
| | | | | | |
| | | | | | |
| | | | | | |
| | | | | | |
| | | | | | |
| | | | | | |
| | | | | | |
| | | | | | |
| | | | | | |
| | | | | | |
| | | | | | |
| | | | | | |
| | | | | | |
| | | | | | |
| | | | | | |
| | | | | | |
| | | | | | |
| | | | | | |
| | | | | | |
| | | | | | |
| | | | | | |
| Ending Balance | | | | | |

*Indicate category code from **[Form 18]**.
**Source: The National Council for Agricultural Education, Decisions & Dollars, 1995

## FORM 20 INCOME AND EXPENSE SUMMARY*

**Enterprise:** _____

Add all income and expense items with the same code and record on this page. Transfer the totals to the statement of cash flows worksheet and the statement of cash flows **[Forms 25 & 26]**.

Note: For analysis purposes, the previous third & fourth level categories (e.g., 2.1.1, 2.1.4.1, etc.) have been combined to the larger categories listed below. The third & fourth level categories needed for financial analysis are calculated on Form 21.

1  cash withdrawals for family living: .......................................................................................... (35) $ _____ E

2  cash paid for operating activities:
   2.1  cash paid for operating activities: feeder animal, feed purchases,
      operating and interest expenses, and hedging account deposits ................................... $ _____ E
   2.2  cash expenses paid in non business operations ....................................................... $ _____ E
   2.3  income and social security taxes paid in cash........................................................ $ _____ E
   2.4  extraordinary items paid in cash ......................................................................... $ _____ E
   2.5  cash withdrawal for family living ......................................................................... $ _____ E
   2.6  cash withdrawals for investment into personal assets .............................................. $ _____ E

3  cash received from operating activities:
   3.1  cash received from business operations (all cash sales from income statement) ............ $ _____ R
   3.2  cash received from non business income and operations........................................... $ _____ R
   3.3  extraordinary items received in cash .................................................................... $ _____ R

4  cash paid for investing activities:
   4.1  breeding & dairy livestock.................................................................................. $ _____ E
   4.2  machinery & equipment ..................................................................................... $ _____ E
   4.3  business real estate; other business assets ........................................................... $ _____ E
   4.4  capital leased assets......................................................................................... $ _____ E
   4.5  bonds & securities; investments in other entities; other non business assets .............. $ _____ R

5  cash received from investing activities:
   5.1  raised breeding & dairy livestock (not capitalized & not depreciated)........................ $ _____ R
   5.2  purchased & raised breeding & dairy livestock (capitalized & depreciated).................. $ _____ R
   5.3  machinery & equipment ..................................................................................... $ _____ R
   5.4  business real estate; other business assets ........................................................... $ _____ R
   5.5  bonds & securities; investments in other entities; other non business assets .............. $ _____ R

6  distribution of dividends, capital, or gifts:
   6.1  cash portion .................................................................................................... $ _____ R/E
   6.2  property portion ............................................................................................... $ _____ R/E

7  cash received from equity contributions:.................................................................... $ _____ R
   7.1  gifts & inheritances .......................................................................................... $ _____ R
   7.2  additions to paid-in capital including investments of personal assets
      into the business ............................................................................................... $ _____ R
   7.3  investments of personal assets into the business ................................................... $ _____ R

8  principal paid on term debt & capital leases:
   8.1  term debt principal payments: scheduled payments ............................................... $ _____ E
   8.2  term debt principal payments: unscheduled payments ............................................ $ _____ E
   8.3  principal portion of payments on capital leases ..................................................... $ _____ E

9  operating loans received (including interest paid by loan renewal): ................................ (53) $ _____ R

10  proceeds from term debt (term debt financing; loans received):.................................... (45) $ _____ R

11  operating debt principal payments: .......................................................................... (54) $ _____ E

E - indicates expenses; R - indicates revenue; R/E - indicates either expense or revenue
Transfer values (35), (53), (45), and (54) to statement of cash flows **[Form 26]**.
*Source: The National Council for Agricultural Education, DECISIONS & DOLLARS, 1995

## FORM 21  INCOME AND EXPENSE SUMMARY*

**Enterprise:** _____

**Interest Expense Schedule:**

| | | beginning of period | end of period | adjustments (+ or −) |
|---|---|---|---|---|
| interest paid in cash or by renewal on operating loans & accounts payable:<br>Total of code # 2.1.4.1 = (A) | | | | (33) |
| interest paid in cash or by renewal on term debt & capital leases:<br>Total of code # 2.1.4.2 = (B) | | | | (22) |
| Total interest paid in cash or by renewal (A + B = C) | | | $ | |
| Transfer values for $^{(32a)}$, $^{(16a)}$, $^{(32b)}$, and $^{(16b)}$ from **[Forms 16 & 17]** | | beginning of period | end of period | adjustments (+ or −) |
| accrued interest on current liabilities: | $[^{(32b)} - {}^{(32a)} = {}^{(32)}]$ | (32a) | (32b) | ± (32) |
| accrued interest on non-current liabilities: | $[^{(16b)} - {}^{(16a)} = {}^{(16)}]$ | (16a) | (16b) | ± (16) |
| | $[^{(32a)} + {}^{(16a)} = {}^{(34a)}]$ and $[^{(32b)} + {}^{(16b)} = {}^{(34b)}]$ | (34a) | (34b) | |
| | $[^{(34a)} + {}^{(34b)} = {}^{(34)}]$ | | (34) | |
| Total change in accrued interest payable $[^{(32)} + {}^{(16)} = D]$ | | | $ | |

Transfer values $^{(34a)}$, $^{(34b)}$, and $^{(34)}$ to income statement: accrual **[Form 24]**.
Transfer values $^{(16)}$, $^{(22)}$, and $^{(34b)}$ to financial worksheet **[Form 29]**.

### Gifts & Inheritance Schedule:

| | |
|---|---|
| cash portion of gifts & inheritances:<br>Total of code # 7.1.1 = (A) | (41a) |
| property portion of gifts & inheritances:<br>Total of code # 7.1.2 = (B) | (41b) |
| Total gifts & inheritances (A + B = C) | (41) |

### Additions to Paid-In Capital Schedule:

| | |
|---|---|
| cash portion of additions to paid-in capital including investments of personal assets into the business:<br>Total of code # 7.2.1 = (A) | (42a) |
| property portion of additions to paid-in capital including investments of personal assets into the business:<br>Total of code # 7.2.2 = (B) | (42b) |
| Total additions to paid-in capital (A + B = C) | (42) |

Transfer values $^{(41a)}$ and $^{(42a)}$ to statement of cash flows worksheet **[Form 25]**.
Transfer values $^{(41)}$ and $^{(42)}$ to statement of owner equity **[Form 28]**.
*Source: The National Council for Agricultural Education, DECISIONS & DOLLARS, 1995

## FORM 22    INCOME AND EXPENSE SUMMARY**

Enterprise: _____

| Capital asset account adjustment for sales and transfers of raised breeding and dairy livestock (not fully capitalized and not depreciated) | | | |
|---|---|---|---|
| | beginning inventory Form 16 | ending inventory Form 17 | adjustment (+ or −) |
| value of raised breeding & dairy livestock<br>use column e or g | −$ | +$ | ±$ |
| cash sales of raised breeding & dairy livestock .............................................. | | | +     (48) |
| net capital asset account adjustment (A) | | | ±$ |
| Capital gain (loss) on breeding and dairy livestock (fully capitalized and depreciated) | | | |
| Sales of: | net sales amount | adjusted cost/basis[1]* | gain (loss) (+ or −) |
| purchased & raised breeding & dairy livestock....................... | +$    (49) | −$ | ±$ |
| death loss or other casualty loss................................. | +$ | −$ | ±$ |
| total gain or loss (B) | | | ±$ |
| Capital gain (loss) on machinery, real estate, and other assets | | | |
| Sales of: | net sales amount | adjusted cost/basis[1]* | gain (loss) (+ or −) |
| machinery & equipment.......................................... | +$    (50) | −$ | ±$ |
| business real estate & other business assets ........................ | +$    (51) | −$ | ±$ |
| total gain or loss C | | | ±$ |
| Gain/loss on sale of capital assets (A + B + C = D) | | | ±$    (52) |

[1]*Usually the adjusted cost/basis is the most current value from column e or g on the inventory sheets [Forms 14 & 15].
Transfer values (48), (49), (50), and (51) from statement of cash flows worksheet [Form 25].
Transfer value (52) to income statements [Forms 23 & 25].
**Source: The National Council for Agricultural Education, DECISIONS & DOLLARS, 1995

## FORM 23    INCOME STATEMENT: CASH*

**Enterprise:** _____    **12-month period ending:** _____

| Revenue | Cash |
|---|---|
| Cash revenue from ending balance of the record of income and expenses [Form 19].<br><br>$ | |
| Gross cash revenues (A) | $ (25) |

| Expenses | |
|---|---|
| Cash expenses from ending balance of the record of income and expenses [Form 19].<br><br>$ | |
| Total cash expenses (B) | $ (26) |
| Net cash income (A − B = C) | $ |
| Depreciation expense [Form 15] (D) | $ (27) |
| Net income from operations (C − D = E) | $ (28) |
| Gain/loss on sale of capital assets (F) | $ (52) |
| Net business income (E − F = G) | $ (29) |
| Non business income (H) | $ (30) |
| Net income (G + H = I) | $ |

Transfer value (52) from income & expense summary [Form 22] & value (30) from SCF worksheet [Form 25].
Transfer totals (25), (26), (27), (28), and (30) to financial worksheet [Form 29] & (29) to SOE [Form 28].
*Source: The National Council for Agricultural Education, DECISIONS & DOLLARS, 1995

## FORM 24   INCOME STATEMENT: ACCRUAL*

**Enterprise:** _____      **12-month period ending:** _____

| Revenue | | | | Accrual |
|---|---|---|---|---|
| Cash revenue from ending balance of the record of income and expenses **[Form 19]**. | | | | |
| | | | $ | |
| | | | Gross cash revenues (A) | $ |

| Inventory adjustments **[Forms 14 & 15]** | Inventories | | Difference (End. − Beg.) | |
|---|---|---|---|---|
| | Beg. | End. | | |
| | $ | $ | $ | |
| | | | Total inventory adjustment (B) | $ |
| | | | Gross revenues (A ± B = C) | $   (25) |

| Expenses | | | | |
|---|---|---|---|---|
| Cash expenses from ending balance of the record of income and expenses **[Form 19]**. | | | | |
| income tax expense **[Form 25]** | | | $   (31) | |
| | | | Total cash expenses (D) | $   (26) |
| | | | Depreciation expense **[Form 15]** (E) | $   (27) |

**Noncash expense adjustment**

| Assets **[Forms 16 & 17]** | Accounts | | Difference (End. − Beg.) | |
|---|---|---|---|---|
| | Beg. | End. | | |
| Unused supplies<br>Other | $ | $ | $ | |

| Liabilities **[Forms 16 & 17]** | | | | |
|---|---|---|---|---|
| Accounts payable. . . . . . . . . . . . .<br>Interest expense **[F. 21]**<br>Other . . . . . . . . . . . . . . . . . . . . | $   (34a) | $   (34b) | $   (34) | |
| | | | Total noncash expense adjustments (F) | $ |
| | | | Net income from operations (C − D − E ± F = G) | $   (28) |
| | | | Gain/loss on sale of capital assets (H) | $   (52) |
| | | | Net business income (G ± H = I) | $   (29) |
| | | | Non business income (J) | $   (30) |
| | | | Net income (I + J = K) | $ |

Transfer value (52) from income & expense summary **[Form 22]** & value (30) from SCF worksheet **[Form 25]**.
Transfer totals (25), (26), (27), (28), (30), (31), and (34b) to financial worksheet **[Form 29]** & (29) to SOE **[Form 28]**.
*Source: The National Council for Agricultural Education, DECISIONS & DOLLARS, 1995

**FORM 25   STATEMENT OF CASH FLOWS WORKSHEET\***

**Enterprise:** _____   **For the period** _____

### Cash flows from operating activities

2 cash paid for operating activities:
- 2.1 cash paid for operating activities: feeder animal, feed purchases, operating and interest expenses, and hedging account deposits .................................................. (a) _____
- 2.2 cash expenses paid in non business operations.......................................................... (b) _____
- 2.3 income and social security taxes paid in cash............................................................ (c) _____ (31)
- 2.4 extraordinary items paid in cash............................................................................... (d) _____
- 2.5 cash withdrawal for family living .............................................................................. (e) _____
- 2.6 cash withdrawals for investment into personal assets............................................... (f) _____

      TOTAL cash paid for operating activities [a + b + c + d + e + f = (36)]: _____

3 cash received from operating activities:
- 3.1 cash received from business operations (all cash sales from income statement)............... (a) _____
- 3.2 cash received from non business income and operations................................................ (b) _____ (30)
- 3.3 extraordinary items received in cash .............................................................................. (c) _____

      TOTAL cash received from operating activities [a + b + c = (37)]: _____

### Cash flows from investing activities

4 cash paid for investing activities:
- 4.1 breeding & dairy livestock ......................................................................................... (a) _____
- 4.2 machinery & equipment............................................................................................. (b) _____
- 4.3 business real estate; other business assets.................................................................. (c) _____
- 4.4 capital leased assets .................................................................................................. (d) _____
- 4.5 bonds & securities; investments in other entities; other non business assets .............. (e) _____

      TOTAL cash paid to purchase [a + b + c + d + e = (38)]: _____

5 cash received from investing activities:
- 5.1 raised breeding & dairy livestock (not capitalized & not depreciated)........................... (a) _____ (48)
- 5.2 purchased & raised breeding & dairy livestock (capitalized & depreciated) ................... (b) _____ (49)
- 5.3 machinery & equipment............................................................................................. (c) _____ (50)
- 5.4 business real estate; other business assets.................................................................. (d) _____ (51)
- 5.5 bonds & securities; investments in other entities; other non business assets .............. (e) _____

      TOTAL cash received from investing activities [a + b + c + d + e = (39)]: _____

### Cash flows from financing activities

6 distribution of dividends, capital, or gifts:
- 6.1 cash portion.............................................................................................................. (a) _____ (40a)
- 6.2 property portion......................................................................................................... (b) _____ (40b)

      TOTAL distribution of dividends, capital, or gifts [a + b = (40)]: _____

7 cash received from equity contributions:
- 7.1 cash portion of gifts & inheritances........................................................................... (a) _____ (41a)
- 7.2 cash portion of additions to paid-in capital including investments of personal assets into the business ...................................................... (b) _____ (42a)

      TOTAL (cash portion) received from equity contributions [a + b = (43)]: _____

8 principal paid on term debt & capital leases:
- 8.1 term debt principal payments: scheduled payments...................................................... (a) _____
- 8.2 term debt principal payments: unscheduled payments................................................. (b) _____
- 8.3 principal portion of payments on capital leases........................................................... (c) _____

      TOTAL principal payments on term debt & capital leases [a + b + c = (44)]: _____
      TOTAL principal payments on term debt & capital leases [a + b + c = (44)]: _____
      TOTAL annual scheduled principal payments on term debt & capital leases [a + c = (23)]: _____

Transfer values (41a) and (42a) from income & expense summary [Form 21].
Transfer values (48), (49), (50), and (51) to income & expense summary [Form 22]; value (30) to income statements [Forms 23 or 24];
values (36), (37), (38), (39), (40a), (43), and (44) to the statement of cash flows [Form 26]; value (40) to statement of owner equity [Form 28]
& values (23) and (31) to financial worksheet [Form 29].
\*Source: The National Council for Agricultural Education, DECISIONS & DOLLARS, 1995

## FORM 26 STATEMENT OF CASH FLOWS*

**Enterprise:**                               **For the period** _____

| | | |
|---|---|---|
| **Cash flows from operating activities** | | |
| 1 cash withdrawals for family living: ........................................................ | | (35) |
| 2 cash paid for operating activities:......................................................... | | (36) |
| 3 cash received from operating activities: ............................................... | | (37) |
| other | | |
| Net cash provided by operating activities (A) | $ | |
| **Cash flows from investing activities** | | |
| 4 cash paid for investing activities: ........................................................ | | (38) |
| 5 cash received from investing activities: ............................................... | | (39) |
| other | | |
| Net cash provided by investing activities (B) | $ | |
| **Cash flows from financing activities** | | |
| 6 distribution of dividends, capital, or gifts (cash portion):.................... | | (40a) |
| 7 cash received from equity contributions: ............................................. | | (43) |
| 8 principal paid on term debt & capital leases: ...................................... | | (44) |
| 9 operating loans received (including interest paid by loan renewal):........ | | (53) |
| 10 proceeds from term debt (term debt financing; loans received): ........... | | (45) |
| 11 operating debt principal payments:..................................................... | | (54) |
| other | | |
| Net cash provided by financing activities (C) | $ | |
| Net increase (decrease) in cash and cash equivalents (A + B + C = D) | $ | |
| Cash and cash equivalents at beginning of year (E) | $ | (15) |
| Cash and cash equivalents at end of year (D + E = F) | $ | |

Transfer value (15) from beginning balance sheet statement **[Form 16]**.
Transfer values (35), (45), (53), and (54) from income & expenses summary **[Form 20]**.
Transfer values (36), (37), (38), (39), (40a), (43), and (44) from the statement of cash flows worksheet **[Form 25]**.
Transfer value (35) to statement of owner equity **[Form 28]**.
Transfer values (35), (36), (37), (38), (39), (40a), (43), (44), and (45) to financial worksheet **[Form 29]**.
*Source: The National Council for Agricultural Education, DECISIONS & DOLLARS, 1995

**FORM 27    CASH FLOW STATEMENT\*\***

**Enterprise:** _____    **For the period** _____

| | Total* | Quarter 1 Jan–Mar | Quarter 2 Apr–Jun | Quarter 3 Jul–Sep | Quarter 4 Oct–Dec |
|---|---|---|---|---|---|
| Beginning cash balance | * | | | | |
| | | | | | |
| Total cash available | * | | | | |
| | | | | | |
| Total cash required | | | | | |
| Cash position before savings and borrowing | * | | | | |
| | | | | | |
| Ending cash balance | * | | | | |

*Amounts in the "Total" column for "Beginning cash balance," "Total cash available," "Cash position before savings and borrowing," and "Ending cash balance" do not equal the sum of the amounts in the four quarters because the ending cash balance for a previous period is carried forward to be the beginning cash balance for the next period. All other amounts in the "Total" column are the sum of the four quarters.
**Source: The National Council for Agricultural Education, DECISIONS & DOLLARS, 1995

## FORM 28 STATEMENT OF OWNER EQUITY*

**Enterprise:**                      **For 12-month period ending:** _____

1. Beginning owner equity[11] ....................................................................(A)   $

2. Net income (accrual)[29] ........................................................................(B)   $ + _____

3. Gifts & inheritances[41] .........................................................................(C)   $ + _____

4. Additions to paid-in capital including investments of
   personal assets into the business[42] ..................................................(D)   $ + _____

5. Distributions of dividends, capital or gifts made (cash
   or property)[40] ......................................................................................(E)   $ − _____

6. Total available ..................................................................(A + B + C + D − E = F)   $ = _____

7. Withdrawals for family living, gifts made, and income and
   Social Security taxes[35] ......................................................................(G)   $ − _____

8. Ending owner equity............................................................. (F − G = H)   $ = _____

Compare H with Owner Equity[14] on Ending Balance Sheet **[Form 17]**.
If different, explain discrepancies here:

Transfer value [11] from **[Form 16]**.
Transfer values [41] and [42] from **[Form 21]**.
Transfer value [29] from **[Form 24]**.
Transfer value [40] from **[Form 25]**.
Transfer value [35] from **[Form 26]**.
*Source: The National Council for Agricultural Education, DECISIONS & DOLLARS, 1995

**FORM 29   FINANCIAL WORKSHEET\***

**Enterprise:** _____

The following values are needed when completing the financial ratios on Forms.

total assets—from balance sheet (BS) **[Form 17]** . . . . . . . . . . . . . . . . . . . . . . . . . . . . . . . . . . . . . (12) _____

total liabilities—from BS **[Form 17]** . . . . . . . . . . . . . . . . . . . . . . . . . . . . . . . . . . . . . . . . . . . . . . . (13) _____

owner equity—from BS **[Form 17]** . . . . . . . . . . . . . . . . . . . . . . . . . . . . . . . . . . . . . . . . . . . . . . . . . (14) _____

beginning cash & cash equivalents—from BS **[Form 16]** . . . . . . . . . . . . . . . . . . . . . . . . . . . . . . . (15) _____

total current assets—from BS **[Form 17]** . . . . . . . . . . . . . . . . . . . . . . . . . . . . . . . . . . . . . . . . . . . . (17) _____

total current liabilities—from BS **[Form 17]** . . . . . . . . . . . . . . . . . . . . . . . . . . . . . . . . . . . . . . . . . (18) _____

average total assets—from BS **[Forms 16 & 17]**
( _____ (9) + _____ (12) ) ÷ 2 = . . . . . . . . . . . . . . . . . . . . . . . . . . . . . . . . . . . . (19) _____

average total liabilities—from BS **[Forms 16 & 17]**
( _____ (10) + _____ (13) ) ÷ 2 = . . . . . . . . . . . . . . . . . . . . . . . . . . . . . . . . . (20) _____

average owner equity—from BS **[Forms 16 & 17]**
( _____ (11) + _____ (14) ) ÷ 2 = . . . . . . . . . . . . . . . . . . . . . . . . . . . . . . . . . (21) _____

annual scheduled principal & interest payments on term debt & capital
leases **[Forms 21 & 25]** ( _____ (22) + _____ (23) ) = . . . . . . . . . . . . . . . . . . . . (24) _____

gross revenue—from income statement (IS) **[Form 24]** . . . . . . . . . . . . . . . . . . . . . . . . . . . . . . . (25) _____

total cash expense—from IS **[Form 24]** . . . . . . . . . . . . . . . . . . . . . . . . . . . . . . . . . . . . . . . . . . . . (26) _____

depreciation expense—from IS **[Form 24]** . . . . . . . . . . . . . . . . . . . . . . . . . . . . . . . . . . . . . . . . . . (27) _____

net income from operations—from IS **[Form 24]** . . . . . . . . . . . . . . . . . . . . . . . . . . . . . . . . . . . . . (28) _____

non business income—from IS **[Form 24]** . . . . . . . . . . . . . . . . . . . . . . . . . . . . . . . . . . . . . . . . . . . (30) _____

income tax expense—from statement of cash flows (SCF)
worksheet **[Form 25]** . . . . . . . . . . . . . . . . . . . . . . . . . . . . . . . . . . . . . . . . . . . . . . . . . . . . (31) _____

interest expense—from IS **[Form 24]** . . . . . . . . . . . . . . . . . . . . . . . . . . . . . . . . . . . . . . . . . . . . . . (34b) _____

cash withdrawals for family living—from (SCF) **[Form 26]** . . . . . . . . . . . . . . . . . . . . . . . . . . . . . (35) _____

cash paid for operating activities—from SCF **[Form 26]** . . . . . . . . . . . . . . . . . . . . . . . . . . . . . . . (36) _____

cash received from operating activities—from SCF **[Form 26]** . . . . . . . . . . . . . . . . . . . . . . . . . . (37) _____

cash paid for investing activities—from SCF **[Form 26]** . . . . . . . . . . . . . . . . . . . . . . . . . . . . . . . . (38) _____

cash received from investing activities—from SCF **[Form 26]** . . . . . . . . . . . . . . . . . . . . . . . . . . . (39) _____

distribution of dividends, capital or gifts (cash portion)—
from SCF **[Form 26]** . . . . . . . . . . . . . . . . . . . . . . . . . . . . . . . . . . . . . . . . . . . . . . . . . . . . (40a) _____

cash received from equity contributions—from SCF **[Form 26]** . . . . . . . . . . . . . . . . . . . . . . . . . . (43) _____

principal paid on term debt & capital leases—from SCF **[Form 26]** . . . . . . . . . . . . . . . . . . . . . . . (44) _____

proceeds from term debt (term debt financing; loans received)—from
SCF **[Form 26]** . . . . . . . . . . . . . . . . . . . . . . . . . . . . . . . . . . . . . . . . . . . . . . . . . . . . . . . (45) _____

interest on term debt & capital leases—from income & expenses
summary **[Form 21]** ( _____ (16) + _____ (22) ) = . . . . . . . . . . . . . . . . . . . . . . (46) _____

value of unpaid operator & family labor & management: this value is in
addition to the value expended on the income statement. Usually, it is
estimated and it represents the opportunity cost of the resources. . . . . . . . . . . . . . . . . . . (47) _____

\*Source: The National Council for Agricultural Education, DECISIONS & DOLLARS, 1995

## FORM 30 FINANCIAL RATIOS*

**Enterprise:** _____ **Date**

Note: All ratios are computed using end of financial year values. You may calculate a ratio at any point in time, but for comparison purposes it is recommended to use a consistent date.

| Liquidity | |
|---|---|
| 1. Working capital | total current assets (17) .............................................. <br> − total current liabilities (18) ............................................. − _____ <br> working capital = $ _____ |
| 2. Current ratio | total current assets (17) .............................................. _____ <br> ÷ total current liabilities (18) ........................................... ÷ _____ <br> current ratio = |
| 3. Cash flow coverage ratio | beginning cash (15) .................................................. <br> + cash received from operating activities (37) ............................ + <br> + cash received from investing activities (39) ............................. + <br> + proceeds from term debt (45)......................................... + <br> + cash received from equity contributions (43) ........................... + _____ <br> ÷ cash paid for operating activities (36)................................. ÷ <br> + cash paid for investing activities (38) ................................. + <br> + principal paid on term loans and capital leases (44)...................... + <br> + cash portion distribution of dividends, capital, or gifts (40a) .................. + <br> cash flow coverage ratio = |

| Solvency | |
|---|---|
| 4. Debt/asset ratio | total liabilities (13)................................................... _____ <br> ÷ total assets (12) ................................................... ÷ <br> debt/asset ratio = |
| 5. Equity/asset ratio | total equity (14)..................................................... _____ <br> ÷ total assets (12) ................................................... ÷ <br> equity/asset ratio = |
| 6. Debt/equity ratio | total liabilities (13)................................................... _____ <br> ÷ total equity (14).................................................... ÷ <br> debt/equity ratio = |

| Profitability | |
|---|---|
| 7. Return on assets | net income from operation (28) ........................................ <br> + interest expense (34b)............................................... + <br> − value of unpaid operator & family labor & management (47).................. − _____ <br> ÷ average total assets (19)............................................. ÷ <br> return on assets = |
| 8. Return on equity | net income from operation (28) ........................................ <br> − value of unpaid operator & family labor & management (47).................. − _____ <br> ÷ average total equity (21)............................................. ÷ <br> return on equity = |
| 9. Operating profit margin ratio | net income from operation (28) ........................................ <br> + interest expense (34b)............................................... + <br> − value of unpaid operator & family labor & management (47).................. − _____ <br> ÷ gross revenue (25) ................................................. ÷ <br> operating profit margin ratio = |

*Source: The National Council for Agricultural Education, DECISIONS & DOLLARS, 1995

## FORM 30    FINANCIAL RATIOS (*CONTINUED*)

**Enterprise:** _____    **Date**

| Repayment Capacity |
|---|

**10. Term debt & capital lease coverage ratio**

net income from operation (28) . . . . . . . . . . . . . . . . . . . . . . . . . . . . . . . . . . . . . . . . .
+ total non business related income (30) . . . . . . . . . . . . . . . . . . . . . . . . . . . . . . . . . . +
+ depreciation/amortization expense (27) . . . . . . . . . . . . . . . . . . . . . . . . . . . . . . . . . +
+ interest on: term debt & capital leases (46) . . . . . . . . . . . . . . . . . . . . . . . . . . . . . . +
− total income tax expense (31) . . . . . . . . . . . . . . . . . . . . . . . . . . . . . . . . . . . . . . . −
− withdrawals for family living (35) . . . . . . . . . . . . . . . . . . . . . . . . . . . . . . . . . . . . − _____
÷ annual scheduled principal and interest payments on term
       debt & on capital leases (24) . . . . . . . . . . . . . . . . . . . . . . . . . . . . . . . . . . . . . ÷

term debt & capital lease coverage ratio =

**11. Capital replacement & term debt repayment margin**

net income from operation (28) . . . . . . . . . . . . . . . . . . . . . . . . . . . . . . . . . . . . . . . .
+ total non business related income (30) . . . . . . . . . . . . . . . . . . . . . . . . . . . . . . . . . . +
+ depreciation/amortization expense (27) . . . . . . . . . . . . . . . . . . . . . . . . . . . . . . . . . +
− total income tax expense (31) . . . . . . . . . . . . . . . . . . . . . . . . . . . . . . . . . . . . . . . −
− withdrawals for family living (35) . . . . . . . . . . . . . . . . . . . . . . . . . . . . . . . . . . . . − _____
= capital replacement & term debt repayment capacity . . . . . . . . . . . . . . . . . . . . . . =
− principal paid on term debt and capital leases (44) . . . . . . . . . . . . . . . . . . . . . . . . −

capital replacement and term debt repayment margin = $

| Financial Efficiency |
|---|

**12. Asset turnover ratio**

gross revenue (25) . . . . . . . . . . . . . . . . . . . . . . . . . . . . . . . . . . . . . . . . . . . . . . . _____
÷ average total assets (19) . . . . . . . . . . . . . . . . . . . . . . . . . . . . . . . . . . . . . . . . . ÷

asset turnover ratio =

**13. Operating expenses ratio**

total cash expense (26) . . . . . . . . . . . . . . . . . . . . . . . . . . . . . . . . . . . . . . . . . . . .
− depreciation expense (27) . . . . . . . . . . . . . . . . . . . . . . . . . . . . . . . . . . . . . . . . − _____
÷ gross revenue (25) . . . . . . . . . . . . . . . . . . . . . . . . . . . . . . . . . . . . . . . . . . . . . ÷

operating expenses ratio =

**14. Depreciation expense ratio**

depreciation expense (27) . . . . . . . . . . . . . . . . . . . . . . . . . . . . . . . . . . . . . . . . . . _____
÷ gross revenue . . . . . . . . . . . . . . . . . . . . . . . . . . . . . . . . . . . . . . . . . . . . . . . . ÷

depreciation expense ratio =

**15. Interest expense ratio**

total interest expense (34b) . . . . . . . . . . . . . . . . . . . . . . . . . . . . . . . . . . . . . . . . . _____
÷ gross revenue (25) . . . . . . . . . . . . . . . . . . . . . . . . . . . . . . . . . . . . . . . . . . . . . ÷

interest expense ratio =

**16. Net income from operations ratio**

net income from operations (28) . . . . . . . . . . . . . . . . . . . . . . . . . . . . . . . . . . . . . . _____
÷ gross revenue (25) . . . . . . . . . . . . . . . . . . . . . . . . . . . . . . . . . . . . . . . . . . . . . ÷

net income from operations ratio =

## FORM 31 TREND ANALYSES***

**Enterprise:**

| Item | For Month and Year Ending | | | | |
|------|---|---|---|---|---|
| **Financial Measures*** | | | | | |
| Total assets—from Form 17[12] | | | | | |
| Total liabilities—from Form 17[13] | | | | | |
| Owner equity—from Form 17[14] | | | | | |
| Gross revenue—from Form 24 (C) | | | | | |
| Total cash expense—from Form 24 (D) | | | | | |
| Net income from operations—from Form 24 (G) | | | | | |
| Net business income—from Form 24 (I) | | | | | |
| **Liquidity**** | | | | | |
| 1.  Working capital | | | | | |
| 2.  Current ratio | | | | | |
| 3.  Cash flow coverage ratio | | | | | |
| **Solvency**** | | | | | |
| 4.  Debt/asset ratio | | | | | |
| 5.  Equity/asset ratio | | | | | |
| 6.  Debt/equity ratio | | | | | |
| **Profitability**** | | | | | |
| 7.  Return on assets | | | | | |
| 8.  Return on equity | | | | | |
| 9.  Operating profit margin ratio | | | | | |
| **Repayment Capacity**** | | | | | |
| 10.  Term debt and capital lease coverage ratio | | | | | |
| 11.  Capital replacement & term debt repayment margin | | | | | |
| **Financial Efficiency**** | | | | | |
| 12.  Asset turnover ratio | | | | | |
| 13.  Operating expenses ratio | | | | | |
| 14.  Depreciation expense ratio | | | | | |
| 15.  Interest expense ratio | | | | | |
| 16.  Net income from operations ratio | | | | | |

*Note: ratios are generally calculated at the end of the financial year.
**Transfer values from financial ratios [**Forms 30 & 31**].
***Source: The National Council for Agricultural Education, DECISIONS & DOLLARS, 1995

## FORM 32    FFA DEGREES AND PARTICIPATION*

### Degrees Held in FFA

| Degree | Year Received |
| --- | --- |
| Greenhand | |
| Chapter | |
| State | |
| American | |
| | |

### Offices Held/Major Committees Chaired in FFA

| Office and/or Major Committee | School Year | Indicate Level | | | |
| --- | --- | --- | --- | --- | --- |
| | | Greenhand | Chapter | District | State |
| | | | | | |
| | | | | | |
| | | | | | |
| | | | | | |
| | | | | | |

### Participation in FFA Activities

| School Year | Leadership events, judging public speaking, banquets, conferences, etc. . . . . (specify) | Indicate Level | | | | Indicate level of participation and/or responsibility |
| --- | --- | --- | --- | --- | --- | --- |
| | | Chapter | District | State | National | |
| | | | | | | |
| | | | | | | |
| | | | | | | |
| | | | | | | |
| | | | | | | |
| | | | | | | |
| | | | | | | |
| | | | | | | |
| | | | | | | |
| | | | | | | |
| | | | | | | |
| | | | | | | |
| | | | | | | |
| | | | | | | |
| | | | | | | |
| | | | | | | |
| | | | | | | |
| | | | | | | |

*Source: The National Council for Agricultural Education, Decisions & Dollars, 1995

**FORM 33 PARTICIPATION***

## Participation in Other School Activities

| School Year | Athletics, school clubs and organizations, etc. . . . (specify) | Indicate Level | | | | Indicate level of participation and/or responsibility |
| | | Chapter | District | State | National | |
|---|---|---|---|---|---|---|
| | | | | | | |
| | | | | | | |
| | | | | | | |
| | | | | | | |
| | | | | | | |
| | | | | | | |
| | | | | | | |
| | | | | | | |
| | | | | | | |
| | | | | | | |
| | | | | | | |
| | | | | | | |
| | | | | | | |
| | | | | | | |
| | | | | | | |
| | | | | | | |

## Participation in Community Activities

| Group/Activity: community sports, youth organizations, church, etc. . . . (specify) | School Year | Indicate level of participation and/or responsibility |
|---|---|---|
| | | |
| | | |
| | | |
| | | |
| | | |
| | | |
| | | |
| | | |
| | | |

*Source: The National Council for Agricultural Education, DECISIONS & DOLLARS, 1995

## FORM 34    AWARDS AND HONORS*

**FFA Awards and Honors**

| Awards and Honors | School Year | Indicate Level | | | |
|---|---|---|---|---|---|
| | | Chapter | District | State | National |
| | | | | | |
| | | | | | |
| | | | | | |
| | | | | | |
| | | | | | |
| | | | | | |
| | | | | | |
| | | | | | |
| | | | | | |
| | | | | | |
| | | | | | |

**Other Awards and Honors**

| Awards and Honors | School Year | Indicate Level | | | |
|---|---|---|---|---|---|
| | | Local | Area | State | Other |
| | | | | | |
| | | | | | |
| | | | | | |
| | | | | | |
| | | | | | |
| | | | | | |
| | | | | | |
| | | | | | |
| | | | | | |
| | | | | | |
| | | | | | |

*Source: The National Council for Agricultural Education, DECISIONS & DOLLARS, 1995

## FORM 35    THE DIARY*

Enter here in chronological order events that are closely related to your SE program that do not belong on the previous pages (e.g., unusual weather, innovative activities, unusual yields, unexpected happenings, SE historical progress, etc.).

| Date | Event, Activity, etc. |
|------|------------------------|
|      |                        |
|      |                        |
|      |                        |
|      |                        |
|      |                        |
|      |                        |
|      |                        |
|      |                        |
|      |                        |
|      |                        |
|      |                        |
|      |                        |
|      |                        |
|      |                        |
|      |                        |
|      |                        |
|      |                        |
|      |                        |
|      |                        |
|      |                        |

*Source: The National Council for Agricultural Education, DECISIONS & DOLLARS, 1995

## FORM 36    SUPERVISION OF STUDENT'S EXPERIENCE PROGRAM*

| Date | Evaluation | | | | Observations and/or Recommendations |
|------|----------|----------|--------------|---------------|--------------------------------------|
|      | Neatness | Accuracy | Completeness | Overall Grade |                                      |
|      |          |          |              |               |                                      |
|      |          |          |              |               |                                      |
|      |          |          |              |               |                                      |
|      |          |          |              |               |                                      |
|      |          |          |              |               |                                      |
|      |          |          |              |               |                                      |
|      |          |          |              |               |                                      |
|      |          |          |              |               |                                      |
|      |          |          |              |               |                                      |
|      |          |          |              |               |                                      |
|      |          |          |              |               |                                      |
|      |          |          |              |               |                                      |
|      |          |          |              |               |                                      |
|      |          |          |              |               |                                      |
|      |          |          |              |               |                                      |
|      |          |          |              |               |                                      |
|      |          |          |              |               |                                      |
|      |          |          |              |               |                                      |

*Source: The National Council for Agricultural Education, DECISIONS & DOLLARS, 1995

# GLOSSARY

**Accounts Payable**   (Ch. 5) Unpaid bills or debts.

**Accrual Accounting**   (Ch. 5) Income is reported in the year earned and expenses are deducted or capitalized in the year incurred. Method of accounting used to compute earnings to be reported to the Internal Revenue Service. Includes increases and decreases in inventory; annual income and expenses; and complete inventory. *See also* cash accounting.

**Accrued Interest**   (Ch. 5) Amount of interest that would be due if the note were paid off on the date the statement is prepared. Interest built up over time.

**Accrued Taxes**   (Ch. 5) Amount of tax liability that would be due if the taxes were paid on the date the statement is prepared. Taxes built up over time.

**Acquisition**   (Ch. 3) Purchase (or receipt) of goods.

**ACRS**   (Ch. 3) Accelerated cost recovery system, a depreciation method in which more depreciation is charged to the early years of the life of the asset.

**Additions to Paid-in Capital**   (Ch. 7) Investments from personal accounts, stocks, or other nonbusiness sources.

**Add-on Interest**   Interest added onto a repayment schedule:

1. Interest calculation
2. Interest added to the amount of the loan
3. Calculation of monthly payment
4. Interest charged over life on loan, therefore APR is higher:
   Stelson equation (to determine actual APR):

$$R = 2 \ C/L \ (P + A) \times 100$$

where
$R$ = APR
$C$ = total interest cost or finance charges
$L$ = length of loan in years
$P$ = beginning principal of the loan
$A$ = amount of each periodic payment

**Amortized**   (Ch. 10) Principal payments plus periodic interest on loans. Amortized loans are in the form of equal principal payments or equal total payments.

**Annual Percentage Rate (APR)**   (Chs. 2, 10) Actual interest rate expressed on an annual basis. *Also known as* effective rate.

**Annuity**   (Ch. 2) A stream of payments or dividends over time that are earned from an investment.

**Appreciable Asset**   Asset that increases in value over time.

**Asset**   (Chs. 3, 4) Anything of value owned by a business or individual.

**ATM**   Automatic teller machine, computer that conducts bank transaction 24 hours a day (subject to the individual bank's policy and service level).

**Audit**   (Ch. 11) Complete examination of financial records.

**Balance Sheet**   (Chs. 2, 4, 5, 6, 7) Financial statement of equity for an individual/business for a specific point in time. List of assets (current or non-current) and liabilities (current and non-current) of an individual or business. *See also* net worth statement.

assets − liabilities = owner equity/net worth/net gain or loss

**Bond**   (Ch. 2) Interest-bearing certificate of debt to bondholder.

**Budget**   (Ch. 9) Formal written or unwritten plan that projects the use of assets for a future time.

**Capital**   (Chs. 3, 4, 6, 10) Cash or liquid savings, machinery, livestock, buildings, and other assets that have a useful life of more than one year and that meet a specified minimum value.

**Capital Gain**   (Ch. 3) Income gain that results from selling a capital asset for more than its adjusted basis (cost less depreciation).

**Cash Accounting**   (Chs. 3, 5) Method of accounting in which income is credited to the year it was received and expenses are deducted in the year they are paid. *See also* accrual accounting.

**Cash Flow**   (Chs. 6, 9) Receipts and disbursements in the business; determines the cash flow budget.

**Cash Flow Budget**   (Ch. 9) Statement of projected expenses and receipts associated with a particular farm or business plan.

**Cash Flow Statement**   (Ch. 6) Projected business payments and receipts associated without particular a business plan.

**Certificate of Deposit**   (Ch. 2) Special form of savings that requires a large amount of money to be invested and left in the account for a certain period of time.

**Check**   Order written by an account holder directing a bank to pay out money from an account.

**Checkbook**   Bound book containing blank checks and check stubs or a check register.

**Checking Account**   (Ch. 2) Bank account against which a depositor may write checks.

**Check Register**   (Ch. 2) Separate form on which the account holder keeps a record of deposits and checks. *Also known as* checkbook ledger.

**Collateral**   (Chs. 2, 3, 10) Assets pledged or mortgaged to secure the repayment of a loan.

**Compounding**   (Chs. 2, 10) Interest received from an investment. Interest is added to the principal and interest is paid again on the total sum. It is a method of calculation that can be used to determine the value.

**Consumable Supplies**   (Ch. 3) Supplies and material used in production.

**Contingent Liabilities**   Amount of money a business would owe if a particular event occurred (e.g., deferred taxes on the sale of current assets, taxes on capital gains, depreciation recapture).

**Cost Behavior**   (Ch. 9) How a cost responds to changes in production volume.

**Credit**   (Chs. 2, 3, 9) Means of obtaining goods or services now by promising to pay at a later date.

**Credit Card**   Card used to obtain goods and services on credit.

**Credit References**   Firms or individuals who have provided credit to someone in the past and can give information on the individual's credit record.

**Credit Union**   (Ch. 2) Cooperative association that accepts savings deposits and makes small loans to its members.

**Current Asset**   (Ch. 4) Asset that is used or sold within the year.

**Current Credit**   Credits that are payable within the year.

**Current Liability**   (Ch. 4) A liability that is payable within the year.

**Current Ratio**   (Ch. 4) Extent to which current assets would cover current liabilities. It is computed by dividing current assets by current liabilities.

**Debt**   (Chs. 4, 8) The liabilities against a person or business.

**Debt-to-Asset Ratio**   (Ch. 4) Ratio computed by dividing total liabilities by total assets. It measures the proportion of total assets owed to creditors.

**Debt-to-Equity Ratio**   (Ch. 4) Ratio that reflects the extent to which debt capital is being combined with equity capital. Divide total liabilities by owner equity. *Also called* leverage ratio.

**Demand**   (Ch. 1) Relationship between price and quantity that illustrates how much consumers are willing and able to purchase at various prices.

**Deposit**   Money placed in a bank account.

**Deposit Slip**   Form that accompanies a deposit and shows the items deposited. *Also known as* deposit ticket.

**Depreciable Assets**   (Ch. 3) Assets that are vital to the business and that lose value over time to wear or obsolescence.

**Depreciation**   (Chs. 3, 11) Decrease in value of business assets caused by wear and obsolescence.

**Depreciation Methods**  (Ch. 3) Formula for depreciating assets (e.g., cost minus depreciation, whether by straight-line, accelerated cost recovery system, or modified accelerated cost recovery system).

**Depreciation System**  A system of accounting that aims to distribute the cost or other basic value of tangible capital assets, less salvage value, over the estimated life of the unit (that may be a group of assets) in a systematic and rational manner.

**Diminishing Marginal Utility**  (Ch. 1) When a product or service is consumed, a time comes when the utility (or pleasure or benefit) of consuming that product or service declines.

**Direct Deposit**  (Ch. 2) Automatic deposit of paychecks into an account.

**Disbursement**  (Ch. 2, 6) Funds paid out.

**Discounting**  (Ch. 2) Present value of a future sum accounting for the rate of compound interest. Procedure used with U.S. Treasury bills that permits buyers to purchase them at a discount and ensures a specific increase in value.

**Discount Interest**  Interest is deducted from the principal at the time the loan is made:

$$R = d/L - d \times 100$$

where
$R$ = APR
$d$ = discount or amount of interest
$L$ = original amount borrowed and to be repaid

**Discount Rate**  Interest rate offered on loan funds to member banks by the Federal Reserve System. This rate is almost always lower than rates offered by banks to customers. *Also known as* prime rate.

**Dividend**  (Ch. 2) Return on investment.

**Economic Opportunities**  (Ch. 1) Possible business prospects available in today's global economy.

**Economic Principles**  (Ch. 1) Basic premises that explain the free enterprise system.

**Economics**  (Ch. 1) Science of allocating and utilizing scarce resources of land, labor, capital, and management in the most optimum manner among different and competing choices to satisfy human wants.

**Enterprise Budget**  (Ch. 9) Full income statement presented for one segment or part of a business (e.g., production of a specific item).

**Equal Marginal Principle**  (Ch. 10) Input should be allocated among alternative uses in such a way that the marginal value products of the last unit are equal in all uses.

**Equilibrium Price and Quantity**  (Ch. 1) Point at which the supply and demand schedules intersect and at which the quantity offered for sale matches exactly the quantity that consumers are willing to purchase.

**Equity**  (Chs. 4, 6) The net ownership of a business. It is the difference between the assets and liabilities of an individual or business, as shown on the balance sheet or financial statement.

assets − liabilities = owner equity

**Equity Capital**  (Ch. 10) Existing capital investment.

**Equity-to-Asset Ratio**  (Ch. 4) Proportion of total assets financed by owner equity capital. Divide owner equity by total assets. *Sometimes also called* percent ownership ratio.

**Estate Tax**  (Ch. 3) Tax on the total amount of property left by a person who dies.

**Exemptions**  (Ch. 11) Deductions from income tax responsibility.

**Expenditure**  (Ch. 6) Payment or expense incurred.

**Expense**  (Chs. 5, 9) Cost of goods or services involved in producing a product or service.

**Feasibility**  (Ch. 4) Capability to execute a plan successfully. It is determined by measuring a business's liquidity and repayment capacity. The statement of cash flows is closely related to feasibility.

**FICA**  (Ch. 11) Federal Insurance Contribution Act, the tax withheld from an employee's paycheck and matched by the employer; the amount is invested in a trust account by the Social Security Administration for a time when the employee no longer works. *Also known as* social security tax or retirement tax.

**Financial Efficiency**  (Chs. 4, 8) A measurement of the degree of efficiency in using labor, management, and capital.

**Financial Statement**   (Ch. 3) A statement that lists the assets and liabilities of a business at a specific time.

**Fiscal Year**   (Ch. 3) Any accounting period that consists of 12 successive calendar months, 52 weeks, or 13 4-week periods. Most businesses use January 1 through December 31.

**Fixed Costs**   (Ch. 9) Costs that remain mostly constant as production changes within a relevant range.

**Free Enterprise**   (Ch. 1) System in which economic agents are free to own property and engage in commercial transactions.

**Future Value**   (Ch. 2) Value (FV) of an investment after it has been compounded with interest for a specified period of time.

$$FV = P(1 + I)^n$$

where
$P$ = principal
$I$ = interest
$n$ = years

**GAAP**   (Ch. 6) Generally Accepted Accounting Principles; accounting guidelines used as standards in the financial world.

**Gift Tax**   (Ch. 11) Tax imposed on property or items received, not earned or paid for.

**Goals**   (Ch. 9) Definitive statement that identifies activities, enterprises, or other achievable levels that you aspire toward.

**Government Policy**   (Ch. 1) United States Department of Agriculture policies that are designed to regulate agribusinesses.

**Gross Margin**   Estimated income above variable costs (total income less total variable costs).

**Gross Pay**   (Ch. 11) Total amount of pay that is due an employee before any withholding.

**Hedging**   (Ch. 1) To offset risk in the market, hedgers use the forward or futures market to guarantee a transaction or price and thereby reduce exchange risk.

**Human Resources**   (Ch. 1) People who work in agribusinesses. Labor used as an input into the production process.

**I-9**   (Ch. 11) Form that documents eligibility to work in the United States. Proof of eligibility is required of every person for any job.

**Income**   (Chs. 5, 6) Payment received for goods or services; can be cash or noncash.

**Income Statement:**   (Chs. 3, 4, 5, 6, 7) Summary of revenues and expenses over a period of time.

**Income Statement Budgets**   (Ch. 9) Three types of budgets can use the income statement format: partial, enterprise, and whole business.

**Income Tax**   (Ch. 11) Tax levied on wages and salaries to finance government activities.

**Inflation**   (Ch. 10) Prices go up; the value of money goes down.

**Inheritance Tax**   Tax on cash or property inherited by an individual.

**Insufficient Funds**   Overdraft not covered by the bank, with the check returned to the depositor.

**Interest**   (Chs. 2, 10) Amount of funds paid to the lender for the use of money.

**International Markets**   (Ch. 1) All markets outside the borders of the United States.

**Inventory**   (Chs. 3, 9) Listing and valuation of all business assets.

**Investment**   (Chs. 2, 6) The outlay of money, usually for income or profit.

**IRS**   (Ch. 11) Internal Revenue Service, the tax-collecting agency for the federal government.

**Leverage**   (Ch. 10) Measure of the extent to which debt capital is being combined with equity capital.

**Liability**   (Chs. 2, 4) Money, goods, and/or services that are owed.

**Liquidity**   (Chs. 4, 6, 8) Consisting of or capable of ready conversion to cash. Measures the ability of a business to meet financial obligations as they come due without disrupting normal ongoing operations.

**MACRS**   (Ch. 3) Modified accelerated cost recovery system, the cost of assets recovered over 3, 5, 7, 10, 15, 20, 27.5, or 31.5 years, depending on the type of property and the statute for that type of property. For the method of depreciation, *see* ACRS.

**Management** (Ch. 1) Process of accomplishing activities associated with the goals of the agribusiness.

**Marginal Input Cost (MIC)** (Ch. 10) Cost of additional input.

**Marginal Value Product (MVP)** (Ch. 10) Additional revenue received from additional input.

**Market** (Ch. 1) Interaction between supply and demand to determine the market price and corresponding quantity bought and sold.

**Market Value** (Ch. 2) Value of an asset on the open market at the present time.

**Marketing** (Ch. 1) Coordination of agricultural production with consumer demand including the conception, pricing, promotion, and distribution of products and services.

**Marketing Plan** (Ch. 1) Agribusiness's written plan of action that details the time frames and activities for implementing a strategy to market its goods and/or services.

**Maturity** (Ch. 2) Point in time when investments come due and are returned to the investor.

**Money Market Accounts** (Ch. 2) Savings account that pays market rate or better interest and allows access to funds without penalty.

**Municipal Bond** (Ch. 2) Long-term, safe investment into city, state, school district, or other governmental unit or agency.

**Mutual Fund** (Ch. 2) Investment in which investors' money is pooled in various types of securities.

**Net Income** (Chs. 5, 7, 9) Revenue minus expenses. Calculated by matching revenues with the expenses incurred to create those revenues, plus the gain or loss on the normal sales of capital assets. *See also* profit.

**Net Pay** (Ch. 11) Gross pay minus all withholding. *See also* take-home pay.

**Net Worth** (Ch. 2) Net assets of an individual or business; owner's equity. *See also* balance sheet.

**Net Worth Statement** Financial statement of equity for an individual or business for a specific point in time. *See also* balance sheet.

**Non-current Asset** (Ch. 4) An asset that is not sold, converted into cash, or used up within the year.

**Non-current Credit** Credit that needs to be repaid over a period of more than one year.

**Non-current Liability** (Ch. 4) A liability that is not due within the year. Formerly classified as intermediate (1–10 years) and non-current (more than 10 years).

**Non-depreciable Asset** (Ch. 3) An asset that does not depreciate, is not kept for over one year, or is not valued at $100 or more.

**Overdraft** Writing checks for more money than is in one's account.

**Owner Equity** (Chs. 4, 7) Assets minus liabilities.

**Owner Equity Statement** (Ch. 3) Financial equity statement of a business for a specific point in time. *Also called* owner equity. *See also* balance sheet.

**Partial Budget** (Ch. 9) A financial statement containing only income statement items that change.

**Passbook** Account book used to record savings account transactions.

**Pay Stub** (Ch. 11) Portion of the paycheck that the employee must retain for three years. The pay stub includes the amount of gross pay, all withholdings and their amounts, and the amount of net pay. *Also known as* payroll statement.

**Present Value** (Ch. 2) Current value (PV) of an investment after it has been discounted (where $P$ = principal, $I$ = interest, $n$ = years).

$$PV = P\frac{1}{(1 + I)^n}$$

**Principal** (Ch. 2, 10) Amount of money borrowed or invested.

**Profit** (Chs. 2, 5) Difference between income and expenditures; net income. *See also* net income.

**Profitability** (Chs. 4, 8) Measurement of the extent to which a business generates a profit or net income from the use of labor, management, and capital. It is also the ability to retain earnings and owner equity growth. A business can survive, compete, and progress if its profitability is at a successful level. The profitability and financial efficiency of the business are criteria usually found in an income statement.

**Receipt**   (Ch. 6) Document, usually a single sheet of paper, that acknowledges the receiving of goods, money, or services.

**Relevant Range**   (Ch. 9) Range of production in which variable and/or fixed costs' behavior remains the same.

**Repayment Capacity**   (Chs. 4, 8) Measurement of the ability to cover term debt and capital lease payments.

**Resources**   (Ch. 9) Sources of supply or support.

**Revenue**   (Chs. 5, 6, 7, 8, 9) Income to the business from the sale of goods or services or changes in inventory. It consists of the proceeds received or value created from current business operations.

**Risk**   (Chs. 2, 4) The chance that the investor might lose the invested funds. Always present, it varies by degree of probability and is linked closely to feasibility and profitability. There is a risk that there may not be a profit and there is usually a risk that a plan may not be executed as intended. Risk is calculated by solvency ratios from information found in the balance sheet.

**Salvage Value**   (Ch. 3) Value of an asset at the end of its useful life (e.g., value of machinery for parts).

**Savings Account**   (Ch. 2) Account on which interest is paid on funds deposited by the account holder.

**Schedule**   (Ch. 11) Income tax form that itemizes deductions and exemptions.

**Self-Employment Tax**   (Ch. 11) Tax liability paid by people who are self-employed instead of paying FICA taxes.

**Simple Interest**   (Ch. 10) Principal × Time in years × Annual rate of interest.

**Social Security**   (Ch. 11) Trust account managed by the Social Security Administration for the purpose of saving money for people to use when they are no longer able to work. A person is identified with a nine-digit social security number. *Also known as* tax identification number.

**Solvency**   (Chs. 4, 8) Describes a business that is in sound financial condition and that is able to pay all its liabilities.

**Statement of Cash Flows**   (Chs. 4, 5, 6, 7) Arranges information into operating activities, investing activities, and financing activities. The sum of the net cash flows is the change in cash during an operating period.

**Statement of Owner Equity**   (Chs. 4, 5, 7) Firm's check on accuracy at a point in time. It verifies that the base financial statements are in agreement and depicts where changes in owner equity occur.

**Stock**   (Chs. 2, 7) Transferable certificates representing partial ownership in a corporation.

**Straight-line Depreciation**   (Ch. 3) Depreciation method is equal to the original cost less salvage value, divided by the years of useful life.

$$\frac{\text{Original cost of asset} - \text{Salvage value}}{\text{Useful life}}$$

**Supply**   (Ch. 1) Relationship between price and quantity that illustrates how much producers are willing to supply at various prices.

**Systematic Decision-Making Process**   (Ch. 9) Process that includes the following steps: (1) define the problem, (2) collect information, (3) evaluate alternative solutions, (4) make decisions consistent with the goals and objectives, (5) implement a solution, (6) evaluate consequences, and (7) accept the consequences.

**Take-home Pay**   (Ch. 11) Amount a person receives after taxes and other deductions are withheld from his or her earnings. *See also* net income, net pay.

**Tax**   (Chs. 2, 11) Compulsory charge, levied, for the common good, by a federal, state, or local unit of government against income or wealth.

**Tax Bracket**   (Ch. 11) Bracket, based on income, that determines what percentage of income will be paid in taxes and which forms to file as an income tax return.

**Time Value of Money**   (Ch. 2) Amount that money increases or decreases over time, depending on its many alternative uses.

**Transaction**   Business or banking function carried out.

**Treasury Bill**   (Ch. 2) Short-term investment in the United States Treasury; sold to investors at a discount (that is, the cost is lower than the return).

**Treasury Bond**   (Ch. 2) Long-term, interest-bearing investment in the United States Treasury.

**Treasury Notes**   (Ch. 2) Investment in the United States Treasury characterized by a large amount of funds, low risk, and high return.

**Unused Supplies**   (Ch. 5) Supplies and materials used in production that were not consumed during the current year.

**Useful Life**   (Ch. 3) Number of years an asset is expected to be valuable to the business.

**Variable Costs**   (Ch. 9) Costs that change as production changes within a relevant range.

**W-2**   (Ch. 11) Form, completed by the employer, that documents the annual totals of gross pay and all tax withholding for an employee that was reported during the year.

**W-4**   (Ch. 11) Form, filed by the employee with the employer, that documents the number of exemptions claimed. The number of exemptions partly determines the employee's tax withholding.

**Whole Business Budget**   (Ch. 9) Full income statements are presented for all segments within the business (e.g., production of a specific item).

**Withdrawal**   Money taken out of a bank account.

**Working Capital**   (Ch. 4) The amount of funds available after sale of current assets and payment of current liabilities. Subtract current liabilities from current assets.

# INDEX

## A

Accelerated cost recovery system (ACRS), basic principles, 63, 65
Account management, 34–35
Accounts payable
  on balance sheets, 79–80
  in income statements, 106, 122
Accounts receivable
  in balance sheets, 76–77
  as revenue, 110–111
Accrual accounting, income statement and, 100–101, 103–106, 113, 115, 122
Accrued interest
  current liabilities, 80–81
  defined, 122
  as expense, 111–114
  non-current liabilities, 82
Accrued taxes
  defined, 122
  as expense, 111–114
Accuracy
  in income statements, 105
  in inventories, 52–53, 56
Acquisitions, inventory of, 56, 65
Additions to paid-in-capital, statement of owner equity and, 161–162, 167
Add-on interest method, 221–222
Agribusiness
  career opportunities, 16–17
  economic opportunities in, 10
  free enterprise principles, 8–11
  government policies and, 11
  human resources management in, 11
  international markets and competitive environment, 14
  management roles and functions, 2–8
  marketing, role of, 11
  marketing categories, 14–15
  marketing plans, 12–13
  risk management, 15–16
  supervised experiences in, 16
Alternative solutions, decision making and, 4
Amortization, of loans, 220, 226
Annual percentage rate (APR)
  calculation of, 42, 46
  cost of borrowing and, 222, 226
Annuity, as investment, 36, 46
Assets
  balance sheet statement of, 74–78, 94
  change in values of, 162–163
  as collateral, 53
  cost and market valuations of, 90–91
  inventory of, 56–57
  return-on-assets ratio, 178
  solvency assessment and, 72–73
  turnover ratio, 181
  valuation of, 57–58
Audits, taxes and, 236, 241
Automated teller machine (ATM), principles of, 35
Automatic withdrawal, advantages of, 35

## B

Balance sheet
  areas covered by, 72–74
  asset/liability cost and market valuation, 90–92
  beginning balance sheet example, 85
  completed balance sheets, 84–86
  ending balance sheet, 86, 164
  financial statements and, 70–72
  income statement and, 100
  liabilities in, 78–84
  liquidity measures in, 87–88
  for loans and credit, 41–42, 46
  owner equity in, 84
  principles of, 69–70, 94
  purpose of, 69–70, 122
  sample MIS forms for, 247, 267–268
  scheduling for, 69–70
  solvency measures in, 89–90
  statement of cash flow and, 150
  statement of owner equity and, 154, 157, 167
  structure and components of, 74–78
Bank reconciliation statements, checking accounts and, 30–33
Bonds, as investment, 36, 46
Borrowed capital
  costs of, 221–222
  decision-making concerning, 212–213
  loan options, 218–220
  loan sources, 220–221
  needs evaluation, 213–215
  overview, 210–211
  risk evaluation, 215–216
Budget
  components of, 193–196
  purposes of, 190, 205
Business income, net business income, 115–116

Business operations
borrowing and decision-making, 212–213
borrowing costs, 221–222
capital, 210–211
cash paid for, coding of, 141
cash received from, coding for, 142
income/expenses, statement of cash flow and, 134
loan fund sources, 220–221
loan options, 218–220
needs evaluation, 213–215
net income from, 114–116
overview, 210
risk evaluation, 215–216
statement of cash flow and, 136–138
taxes and, 236–237
Business trends, balance sheet identification of, 73

## C

Capital. *See also* borrowed capital
defined, 150, 226
income and expense summaries on sale of, 110
inventory of, 57, 65
profitability assessment and, 71, 94
rationing of, 217
replacement of, 180
statement of cash flow, 134–136
coding for, 142–144
statement of owner equity and changes in, 159–160
working capital, 88
Capital gains
statement of cash flow coding for, 142–143
tax, inventory control and, 54, 65
Capital leases
as non-current liability, 84
principal payments on, statement of cash flow coding for, 143
repayment capacity and, 179–180
Career overview, agribusiness, 45
Cash accounting
defined, 122
income statement and, 53, 100, 103–106, 112, 115
inventory and, 61, 65
Cash flow budget
capital needs assessment, 213–215
components of, 201–202
defined, 193–196, 205
sample MIS form for, 246, 263
Cash flow coverage ratio, 176–177
Cash flow statement, 69–70, 85, 122
base financial statements and, 133–134
basic principles, 128–130, 150
borrowed capital and, 210–211
for business and lenders, 136–138
completion codes for, 139–148
income statement and, 100, 138–139

sample MIS forms for, 248, 281–283
sample of, 130
sources of cash on, 134–136
statement of owner equity and, 159–160
uses of, 131–133
Cash inflow/outflow projection, statement of cash flow, 132
Cash on hand, in balance sheets, 76
Cash sales, as revenue, 106–111
Certificates of deposit, 22, 46
Chattel mortgages, in balance sheets, 83
Checkbook ledger, sample page, 29
Checking accounts
in balance sheets, 76
bank reconciliation statements, 30–33
basic principles, 27–29, 46
check preparation guidelines and samples, 29, 31
sample statement, 29
Check register, 46
Coding system, statement of cash flow, 139–148
Collateral
balance sheet identification of, 73
inventoried assets and, 53, 65
for loans, 42, 46, 218, 226
Competition, marketing and, 14
Compounding
cost of borrowing and, 221–222
interest table for, 223
principles of, 37–39, 46, 227
Consumable supplies, inventory of, 56, 65
Contributed capital, statement of owner equity and, 159–162
Cost behavior, budgeting and, 190, 205
Costs, variable and fixed costs, 190–193
Costs of borrowing, 221–222
Cost valuation, of assets and liabilities, 90–92
Coverage ratios, term debt and capital leases, 179–180
Credit
basic principles of, 39–42, 46, 65
cash flow budget and, 201–202, 205
financial management and, 22
importance of, 40
inventory and application for, 52–53
maintenance of line of credit, 133
sources of, 40–42
Credit cards, 39–40
Credit history, importance of, 39–40
Credit unions, 42, 46
Current assets, defined, 74–76, 94
Current liabilities
accrued interest on, 80–81
on balance sheets, 79–80, 94
Current ratio
on balance sheet, 88, 94
liquidity and, 173–176

## D

Debt
    feasibility assessment and, 71, 94
    financial performance analysis and,
        174–183, 185
    non-current liabilities, 80
    principal payments on, statement of cash
        flow coding for, 143
Debt-to-asset ratio, 89, 94, 177
Debt-to-equity ratio, 89, 94, 177
    capital needs assessment, 214–215
    risk management and, 215–216
Decision making
    agribusiness management, 3–6
    basic principles, 190
    borrowing and, 212–213
    budget tools, 190, 193–196
    cash flow budget, 201–202
    systematic planning and, 196–201
    variable and fixed costs, 190–193
Decision rule, risk management and, 215–216
Deductions
    business deductions, 237
    taxes and, 236
Deferred tax liability, defined, 84
Demand, free enterprise principles and, 8–9
Deposit slips, sample of, 34
Depreciation
    of assets, 58–61, 65
    business deductions and, 237
    defined, 241
    depreciable inventory, sample MIS forms
        for, 247, 265–266
    expense ratio, 181
    methods of, 63, 66
    salvage value and, 59
Diaries, sample MIS forms for, 249
Diminishing marginal utility, defined, 8
Direct deposit, advantages of, 35, 46
Disbursements
    from checking accounts, 27, 46
    defined, 150
    summary of, statement of cash flow
        and, 131
Discounted variable costs, planning and
    decision-making and, 192
Discounting, present value calculations and,
    37–39
Dividends
    investments and, 36, 46
    statement of cash flow coding for, 142, 144
Documentation requirements
    personal records, 236
    taxes and, 232–234

## E

Economic opportunities, in agribusiness, 10
Economics
    agribusiness management and, 8

free enterprise principles and, 8–11
Employee tax withholdings, 234–236
Employer documentation, taxes and,
    232–234
Employer expectations, personal financial
    management, 42–45
Ending balance sheet, sample form, 86, 164
Enterprise budget
    format for, 193–196, 205
    sample MIS form for, 245, 261
Equal marginal principle, 217, 227
Equilibrium price and quantity, free
    enterprise principles and, 10–11
Equity
    on balance sheets, 84, 95
    contributions, statement of cash flow
        coding for, 142
    defined, 150
    return-on-equity ratio, 179
    statement of owner equity and changes
        in, 159–160
Equity capital, 215, 227
Equity-to-asset ratio, 89, 95, 177
Estate planning, inventory and, 53–54
Estate taxes, inventory and, 54, 66
Evaluation process
    capital needs assessment, 213–215
    decision-making and, 190
    decision making and role of, 5
Exemptions. *See also* deductions
    taxes and, 233–234, 241
Expense ratios, 181
Expenses
    in budget, 190
    defined, 122, 150, 205
    on income statements, 100, 111–114
    sample MIS forms for, 248, 269–278
    in statement of cash flow, 134, 141
    summary of, statement of cash flow and,
        131, 141
Expense summary sheets, 106–111
    supervised experiences sample form,
        120–121

## F

Family living, cash withdrawals for,
    statement of cash flow and, 140
Feasibility, balance sheet and indicators of,
    70–71, 95
Federal Insurance Contribution Act (FICA)
    tax, 234–236, 241
Financial activities, in statement of cash
    flow, 136
Financial analysis
    basic principles, 172–173
    financial efficiency, 174, 180–181
    guidelines for, 182–183
    liquidity, 173–177
    main components of, 174–176
    profitability, 174, 178–179

Financial analysis *(continued)*
    repayment capacity, 174, 179–180
    solvency, 174, 177–179
Financial efficiency
    financial performance analysis, 172, 176,
        180–181, 185
    profitability assessment and, 71–72, 95
Financial management
    account management, 34–35
    advantages and disadvantages of, 25–27
    bank statement reconciliation, 30–31
    career overview, 45
    checking accounts, 27–31
    check preparation and completion
        guidelines, 29–30
    credit, 39–41
    employer expectations and, 41–42
    importance of, 23–24
    investments and savings, 35–37
    life skills in, 22–23
    savings accounts, 31–34
    supervised experience in, 43–44
    value of money and, 37–39
Financial obligations, statement of cash flow
    and ability to meet, 133
Financial risk. *See* risk management
Financial statements
    basic principles of, 52, 66
    income statement and, 102–103
    sample MIS form for, 249, 285
    statement of cash flow and, 133–134
    statement of owner equity, 154–155
Fiscal year, inventory scheduling and,
    55–56, 66
Fixed costs, planning and decision-making
    and, 190–193, 205
Formula pricing markets, defined, 14
Form utility, marketing and, 11–12
Free enterprise principles, agribusiness
    management and, 8–11
Funding sources for loans, 220–221
Future Farmers of America (FFA)
    awards and honors form, 291
    degrees and participation form,
        289–290
Future value, defined, 37–39, 46

**G**

Generally Accepted Accounting Principles
    (GAAP)
    defined, 150
    statement of cash flow preparation,
        128–129
Gifts, statement of cash flow coding for,
    142, 144
Goals
    financial goal guidelines, 213
    multiple goal satisfaction guidelines, 217
    planning and decision-making and,
        190, 205

    systematic planning for, 198–201
Government policies, agribusiness and, 11
Gross pay, taxes and, 236, 241
Group disciplined markets, defined, 14

**H**

Hedging, risk management and, 15–16
Historical practices, statement of cash flow
    and evaluation of, 131–132
Human resources, agribusiness and, 11

**I**

Implementation, decision making and role
    of, 4–5
Income
    business taxable income, 236–237
    cash and accrual accounting
        identification of, 103–106
    defined, 122, 150
    net income, 114–116
    sample MIS forms for, 248, 269–278
    statement of cash flow and evaluation of,
        131–132, 134
Income and expenses code sheet
    sample MIS forms for, 247–248, 269–270
    statement of cash flow and, 140
Income statement
    accruals, 158, 280
    balance sheets and, 69, 95
    budget format, 193–196, 205
    cash and accrual accounting, 103–106
    defined, 100, 122, 150
    expenses, 111–114
    financial statements and, 102–103
    importance of, 100–101
    inventory accuracy in, 52–53, 66
    net income, 114–116
    revenue, 106–111
    sample MIS forms for, 248, 279–280
    statement of cash flow and, 100, 138–139
    statement of owner equity and,
        154, 167
    uses of, 101–102
Income summary sheets, 106–111
Income taxes
    basic principles, 232, 241
    filing guidelines, 236–237
Inflation, cost of borrowing and, 222, 227
Information gathering and management,
    decision making and, 3
Insufficient funds, procedures for, 35
Insurance, inventory and, 53
Interest
    accrued interest
        current liabilities, 80–81
        non-current liabilities, 82
    in checking accounts, 27, 46
    expense ratio, 181
    on loans, 219, 227

Internal Revenue Service (IRS), audits by, 236, 241
International markets, competition, 14
Inventory
    accuracy in, 52–53
    adjustments as revenue in, 106–111
    basic principles of, 52, 66, 150
    cash, receivables, and personal assets and liabilities, 62
    contents of, 56–57
    credit requirements and, 53
    depreciable assets, 58–61
    depreciation methods, 63
    estate planning, 53–54
    income statement and, 53
    insurance and, 53
    non-depreciable assets, 57–58
    non-depreciable inventory, 77
    resource inventory, 199–201, 205
    sample MIS forms for, 246–247
    scheduling of, 55–56
    statement of owner equity, 53
    supervised experience with, 63–64
    tax management, 54
    values of, 57–62
Investments
    in balance sheets, 76
    cash paid for, coding of, 142
    financial management and, 22
    as savings, 35–37, 46
    in statement of cash flow, 134–136
I-9 tax form, 233, 241

**L**

Legislated markets, defined, 14
Lenders, statement of cash flow for, 136–138
Leverage
    capital needs assessment, 214–215
    defined, 227
    risk management and, 215–216
Liabilities
    on balance sheets, 78–84, 95
    cost and market valuations of, 90–91
    current liabilities, 79–81
    non-current liabilities, 80–84
Life skills development, financial management and, 22–45
Line of credit, maintenance of, 133
Liquidity
    balance sheet indicators of, 72, 95
    cash flow coverage ratio, 176–177
    current ratio, 173
    defined, 150
    feasibility assessment and, 71
    financial performance analysis, 172–177, 185
    measures of, 87–88
    in savings and investments, 35–37, 46
    statement of cash flow and, 131–133
    working capital, 173

Loans. *See also* credit
    costs of borrowing, 221–222
    criteria and options for, 218–220
    funding sources for, 220–221
    length of, 218
    as non-current liability, 82, 84
    operating loans, statement of cash flow coding for, 144, 146–147
    personal loans, 41–42
    repayment schedules, 219–220
Long-term credit, basic principles of, 39–42

**M**

Management
    agribusiness and role and function of, 2–8
    functions of, 6–7
    tax management, 237–238
Management information systems (MIS)
    basic principles, 244–249
    sample forms, 250–292
Marginal input costs, risk management and, 216, 227
Marginal value product, risk management and, 216, 227
Marketing
    competitive environments and internation markets, 14
    defined, 11
    importance of, 11–12
    plans, principles for, 12–13
    risk management and, 15–16
Markets, categories of, 14–15
Market value
    of assets and liabilities, 90–92
    of investments, 36, 46
Maturity
    certificates of deposit, 33
    of investments, 46
Modified accelerated cost recovery system (MACRS), basic principles, 63, 66
Money market accounts
    financial management and, 22
    as savings accounts, 32, 47
Mortgages, in balance sheets, 83–84
Municipal bonds, as investments, 33, 47
Mutual funds, as investments, 33, 47

**N**

Negative cash flow, investments and assets and, 134–136
Negotiated markets, defined, 15
Net income
    basic principles, 114–116, 122, 205
    budgeting and role of, 195
    income statement and, 100
    operations ratio, 181
    statement of owner equity and, 154, 167
Net pay, defined, 239–241

Net worth, personal financial management and, 25, 47
Non-current assets, defined, 76–79, 95
Non-current liabilities
  accrued interest on, 82
  in balance sheets, 80
  current portion, in balance sheets, 80
  defined, 82–83, 95
Non-depreciable inventory
  in balance sheets, 77
  sample MIS forms for, 247, 264
  valuation of, 57–58, 66
Nonreal estate loans, defined, 218
Notes payable, in balance sheets, 79–80, 83
Notes receivable, in balance sheets, 76–77

## O

Occupational safety, 43
Open markets, defined, 14
Operating expenses ratio, 181
Operating loans
  principal payments, statement of cash flow coding for, 147
  statement of cash flow coding for, 144
Operating profit margin ratio, profitability and, 179
Operational decisions, properties, 5
Organizational decisions, properties, 5
Overdraft protection, advantages of, 35
Owner equity
  balance sheet and, 70, 84, 95
  calculated *vs.* reported, 165–166
  income statement and, 100
  statement of, 52, 66, 154–165, 167, 248, 284

## P

Partial budget
  format for, 194, 205
  sample MIS form for, 246, 262
Passbook savings accounts, 32
Past financial practices, statement of cash flow and evaluation of, 131–132
Payroll calculations, 234–235
  personal records and, 236
Pay stubs, taxes and, 236, 241
Personal assets and liabilities, inventory of, 61
Personal financial management
  advantages and disadvantages, 25
  reasons for, 23–24
  sample forms for, 26–28
Personal loans
  basic principles of, 41–42
  defined, 218
Personal records, taxes and, 236
Place utility, marketing and, 11–12
Planning
  decision-making and, 190

systematic planning, 196–201
Positive cash flow, investments and assets and, 134–136
Possession utility, marketing and, 11–12
Present value, defined, 37–39, 47
Principal
  compounding calculations and, 37–39, 47
  on loans, 219, 227
Problem identification, decision making and, 3
Production levels, fixed costs and, 193
Profit
  income statement and, 100, 122
  in investments, 22, 47
  maximization of, 215–216
Profitability
  balance sheet and indicators of, 70–72, 95
  financial performance analysis, 172, 178–179, 185
Property tax
  as accrued liability, 82
  inventory control and, 54

## R

Ratios
  cash flow coverage, 176–177
  current ratio, 173
  equity-to-asset, 89, 95, 177
  financial analysis using, 182
  operating expenses, 181
  operating profit margins, 179
  return-on-assets, 178
  return-on-equity, 179
  sample MIS forms for, 249, 286–287
Real estate
  loans, defined, 218
  mortgages/contracts, in balance sheets, 84
  tax, as accrued liability, 82
Receipts
  defined, 150
  summary of, statement of cash flow and, 131
Receivables, inventory of, 61
Regulated markets, defined, 15
Relevant ranges, variable and fixed costs, 192–193, 205
Repayment capacity
  feasibility assessment and, 71, 95
  financial performance analysis, 172, 174, 179–180, 185
Repayment schedules for loans, 219–220
Resources
  budget planning and, 190
  inventory, systematic planning using, 199–201, 205
Resumes, sample MIS forms for, 249
Retained earning, statement of owner equity and changes in, 155–159
Return-on-assets ratio, profitability and, 178

Return-on-equity ratio, profitability and, 179
Revenue
  categories for, 109–111
  defined, 122, 150, 186, 205
  income statement and, 100, 106–111
  profitability and, 178–179
  retained earnings changes and, 155–159
  statement of cash flow and evaluation of, 131–132
  statement of owner equity and, 167
Risk-bearing capacity, balance sheet assessment of, 73
Risk management
  balance sheet and indicators of, 70–71, 95
  borrowed capital and, 215–216
  marketing and, 15–16
  in savings and investments, 35–37, 47

S

Safety practices, personal and occupational safety, 43
Salvage value, depreciation and, 59, 66
Savings account
  advantages and disadvantages, 33–34
  in balance sheets, 76
  basic principles of, 31, 47
  financial management and, 22
  types of, 32–33
Security, for loans, 218
Self-employment tax, basic principles, 232, 238, 241
Short-term credit, basic principles of, 39–42
Simple interest calculations, 221, 227
Social Security number, tax eligibility and, 233, 241
Solvency
  balance sheets and, 69, 72, 95
  financial performance analysis, 172, 174, 177–178, 186
  measures of, 89–90
Statement of cash flows, 69–70, 85, 122
  base financial statements and, 133–134
  basic principles, 128–130, 150, 167
  for business and lenders, 136–138
  completion codes for, 139–148
  income statement and, 100, 138–139
  owner equity and, 154–165
  sample MIS forms for, 248, 281–283
  sample of, 129
  sources of cash on, 134–136
  uses of, 131–133
  worksheet, 145, 163
Statement of owner equity, 52–53, 95
  balance sheet and, 70
  basic principles, 154, 167
  calculated vs. reported equity, 165
  financial statements and, 154–155
  income statement and, 100, 123
  purposes of, 154
  retained earnings changes, 155–159

  sample MIS forms for, 248, 284
  total owner equity changes, 155–164
  withdrawals and, 159
Stock
  as investment, 22, 33, 47
  statement of owner equity and, 160–161, 167
Straight-line depreciation, basic principles, 63, 66
Summary of receipts and disbursements, statement of cash flow and, 131
Supervised experiences
  balance sheet management, 92
  business borrowing, 224–225
  diary form, 292
  exploratory experience forms, 244, 252–253
  in financial management, 43–44
  improvement activities forms, 245, 260
  income statement management, 116–119
  inventory management, 63–64
  placement experiences forms, 245, 254–257
  planning and decision-making, 202–203
  program supervision form, 249
  sample record form for, 244, 250–251
  school-based experience forms, 245, 258–259
  statement of cash flow preparation, 148
  student's experiences form, 293
  tax management, 238
Supply, free enterprise principles and, 9–10
Systematic planning, decision-making process, 196–201, 205

T

Take-home pay, taxes and, 234–236, 241
Tax bracket, 236, 241
Taxes
  accrued taxes, 111–114
  business taxes, 236–237
  deferred tax liabilities, 84
  employer's documents, 232–234
  estate taxes, 54
  filing guidelines, 236–237
  income tax returns, 236
  inventory and, 53–55
  management guidelines, 237–238
  overview, 232
  personal financial management and, 23, 33, 47
  schedules for, 236, 241
  self-employment tax, 232, 238
  take-home pay, 234–236
Term debt
  principal payments on, statement of cash flow coding for, 143
  proceeds from, statement of cash flow coding for, 146
  repayment capacity and, 179–180

Time utility, marketing and, 11–12
Time value of money, principles of, 37–39, 47
Treasury bills, bonds, and notes, as investments, 33, 46–47
Trend analyses
    financial performance measurement and, 182–183
    sample MIS forms for, 249, 288
Turnover ratio, assets, 181

## U

Unused supplies
    defined, 123
    as expense, 111–114
Useful life valuation, non-depreciable assets, 57–58, 66
Utilities, in marketing, 11–12

## V

Values, inventory values, 57–63
Variable costs, planning and decision-making and, 190–193, 205

## W

Whole business budget, format for, 194, 205
Withdrawals, statement of owner equity and, 159
Withdrawal slip, sample of, 34
Withholdings, taxes and, 236
Work habits, employer expectations for, 42–43
Working capital
    on balance sheet, 88, 95
    liquidity and, 173
W-2 tax form, 236, 242
W-4 tax form, 233, 242